TED CRAIL

Contemporary Books, Inc.
Chicago

Library of Congress Cataloging in Publication Data

Crail, Ted.
Apetalk & whalespeak.

Includes index.
1. Human-animal communication. I. Title.
QL776.C7 1983 599.059 82-21996
ISBN 0-8092-5527-8 (pbk.)

This edition published by arrangement with J. P. Tarcher, Inc.

Published by Contemporary Books, Inc.
180 North Michigan Avenue, Chicago, Illinois 60601
Manufactured in the United States of America
Library of Congress Catalog Card Number: 82-21996
International Standard Book Number: 0-8092-5527-8

Published simultaneously in Canada by
Beaverbooks, Ltd.
150 Lesmill Road
Don Mills, Ontario M3B 2T5
Canada

To Lee, the artist
To Charr, the photographer
To Gump, the dog
To Dean, the sculptor
To Bobby, the pool-player
To Shari, the lawyer
To Myshkin, the cat
To Gerald, the typewriter spider

. . . great intercommunicating creatures, every one

Contents

Acknowledgments

THE TOLERANCE OF A GREAT MANY PERSONS HAS BEEN tested in the writing of this book. I would like to acknowledge the ingenuity and insight of a number of persons who have contributed directly or indirectly to the manuscript. Belton P. Mouras, president of the Animal Protection Institute of America, introduced me to the subjects of humane education and animal consciousness, helping acquaint me with a vast world movement directed to the scientific study of animal mentality. Belton would be first of all, I think, in saying to a talking animal, "I may disagree with what you have to say but I shall defend to the death your right to say it." Indispensable in the creation of this book in the form in which you see it was Kirsten Grimstad, the editor who has guided me through writing, research, and final organization of the material. The whole project would have foundered without the candor of Penny Patterson, keeper of gorillas and exploratory investigator into the real possibilities of interspecies communication; the exuberance of Jim Nollman, who has sought the means for talking with whales; the thirty-year outpourings of Dr. John Cunningham Lilly on the extraterrestrial mindset of dolphins and whales. Lee, my wife, has contributed as many pounds of effort as a bowhead whale to all that you will find here. More than this, I think the scientist-adventurers who have pursued the search for animal consciousness create a path to the future for all of us and I salute their adventuring spirit.

Preface

A Dog, Seventeen Cats, & Dr. Lilly's Prophecy

ALL ANIMALS ARE NOT CREATED EQUAL. I KNEW THIS from earliest years, but the knowledge arrived under odd circumstances.

When I was six years old, a doctor found that I was dying of Bright's disease. He saw no chance, he told the adults who cared for me, that I could live until Christmas. My home at this time was in the back of a tire shop on the main street of town in Kalispell, Montana. Sending me to Johns Hopkins was out of the question. In any case, those who watched out for me had better ways than that of outwitting the local doctors.

They conceived a plan to keep me from missing Christmas. They gave me Christmas in September. A tree was cut, popcorn balls were popped, loads of presents rounded up. I was photographed playing my new toy saxophone against a September Christmas tree, shimmering with tinsel. I never thought to question why I alone was enjoying Christmas at this unusual time of year. My principal co-celebrator, a black-and-white dog named Gump, didn't question it, either.

For another year or two thereafter, I continued to die under the best possible circumstances. Adults took me to see grizzly bears and glaciers. They wanted me to see it all—everything that made Montana an eye-popping wonder. Teachers held me in their arms to impart the glory of books so I would not die an illiterate. Even though I gradually became totally ambulatory, running outside, making snowballs, adults still gave me whatever I wanted. I wished to marry Shirley Temple and they said I could; they were denying me nothing.

Poor Shirley Temple! Marooned by stardom, she invented an imaginary playmate to tell her secrets to. Marooned by illness—almost exactly her own age—I made Shirley Temple my far-off playmate, but my close-up playmate was Gump. And Gump was better than Shirley's fictional friend because he listened well, he wasn't invisible, and, as we all knew, he was like a dog inspired. "A genius!" said Guy Ludwick, the man who raised me and who taught Gump his tricks.

Whether Gump was a genius or not, he was the best fun I had in the period when it was intended that, since I had to die, I was certainly going to die in a state of rapture. As I inexplicably grew better, we began to play baseball every afternoon in front of the tire shop just after the stores closed. Guy was the pitcher. Gump was the catcher. I was the batter. If I swung and missed, Gump would catch the ball and deliver it to Guy. If I got a lucky hit, Gump suddenly became an outfielder. He would race off after the ball, grab it, and plunge into me with the ball in his mouth before I could reach second base. That was an Out.

Stopped by the curiosity of a ball game at the center of town, the gallery of onlookers which formed found it amazing—not that I could get a hit but that Guy kept yelling, "Tag him, tag him, tag him, get him out!" and the dog seemed to do it. As far as I know, Gump was just trying to catch up and give me the ball, but the spectators accepted him as a shrewd baseball player and loved it when Guy said, "That dog's a great shortstop!"

My lingering illness turned into lingering health. Gump died and the doctor died. Though he tried hard, Guy never found another dog who could learn the same tricks. For me, there were significant leftovers from this childhood experience. I was left with a distrust of those who insist they know "the scientific facts" and are very peremptory about it (I have been declared dying a time or two since and it always seems to be the doctor who dies). Childhood also gave me the view that animals can be taught to do wonderful things—some animals more than others.

How wonderful? This I have never known. I have merely kept my eyes open for it. It is an idea that can be pushed too far—an idea that, like a doctor's prophecies, should be looked at closely and measured against objective reality. There are limits; Gump did not, after all, become a shortstop for the Giants. But just what those limits are may be broader than we imagine. Within my own lifetime, the country doctors began to learn that Bright's disease is not one thing but many things, not all of them fatal, as was once thought. Likewise, there were authorities—and just ordinary folk—who began to learn that those narrow ideas humans have about the capacities of animals can be padded with all sorts of official fictions and spurious science.

We can be too quick, I think, with the belief "Why, that animal's amazing—understands everything I say." But we can also be too slow to admit the reality of intelligent action by an animal because some scientific authority has asserted that "language is species specific to man" —that no animal, under any circumstance, ever talks an intelligible language.

The book you are about to read deals with the search for interspecies communication and therefore the questions of animal intelligence and animal consciousness at an advanced experimental level. Its primary focus is on highly specialized inquiries into the chances for developing some type of reliable two-way conversation with whales, dolphins, gorillas, and possibly other animals. I have been at pains to gather the views of those who explore and experiment in this field, as well as the views of those who believe that misapprehension, even prevarification, is likely to be found among the animal communicators.

The very specialness of these arguments should not intimidate you into discounting the meaningfulness of your own contacts with animals at the household and neighborhood level. Most of us will never have a chance to exchange greetings with a whale while meeting it underwater, but it is no great trick to go among horses and see if they don't become more understandable in the light of a "horse vocabulary" described in Chapter Two that finds separate meanings in a number of horses' most common gestures.

My own curiosity about the animal communication quandary was fueled, at the beginning of the skindiving era, by interviews I had as a reporter with flippered adventurers coming back from the ocean. I remember a champion swimmer who told me that he could read sharks "the way you read a dog" and that he had been able to swim peacefully with a great white shark because he understood its signals. I didn't become truly entangled with trying to fathom animal language, though, until a time came when I found myself with an almost overwhelming tribe of communicating cats.

At the outset the cats were only three, all female. Through the miracle of triple motherhood, three cats became a total of seventeen cats. Seventeen cats is a dream and a nightmare in the art of communication between species.

I discovered, from studying my seventeen cats, each of whom was very different from the other sixteen, that cats make their desires easily known through what the communicators speak of as "gestural language." Cats lead you to their food bowl, lead you to the front door, appear like apparitions in the window when you do not let them in. There are cats who knock at front or back door in a close simulation of human rapping. There are cats who sit on books as if to say, "My turn now—pay attention." Cats are not really at a loss to communicate basic needs and intentions even though their language system is admittedly incomplete.

Cats also put up blank walls—by instinct, by lack of understanding, or through some great fondness for *not* understanding. None of the seventeen cats could be shown why it was not right to tear great gaping holes in our patio screening. I had no communication method to convince them of this, although many were tried. The holes were their keys to freedom on a moment's notice, and they felt they needed this avenue of escape. I understood their message—that they needed greater freedom—but they did not understand mine—that the neighbors would turn into a lynch party if I couldn't keep the cats in control.

Perhaps the most poignant lesson from this period came as a result of our constant admonitions to our daughter, Charr, five years old and a manhandler of cats, that she could not pull, yank, squeeze, chase, and do great belly flops on her favorite female cat and still expect it to love her. Charr ignored this. She never learned. Then one night, Charr woke us twice, wandering into our bedroom, crying, "Mice! Mice!"

"Baby, mice can't harm you," said her mother, "And we don't have any mice, Charr."

In the morning, we found the mice. They were tiny kittens, born in the middle of Charr's bed. The cat that Charr had so ferociously manhandled—instead of going off to a dark corner as cats usually do to have babies—had given birth in her bed. For all the yanks and squeezes, the cat understood who loved her best.

Many humans communicate well enough with animals from the beginning—Charr certainly did—that it is bewildering to be told in later life that communicating with animals cannot advance past a rudimentary stage. It cannot become, for instance, what so many of the interspecies communicators you are about to meet have in mind—a type of human-to-animal and animal-to-human conversation.

It can't? This is a recognizable red flag to many animal lovers, and lately it has seemed an unimaginative conclusion to some professionally trained animal researchers. We will hear from parties on both sides of that controversy as we go along. I still have two cats in my life (times change) and confess that my sympathy and my desire lie with the communicators. Who does not want a porpoise to talk? My brain, though, cannot ignore the skeptics. Although my experiences with cats were multiplied ten thousand times, I began to recognize certain limits to what they could say by gesture and meow. As to porpoises, my mind stayed open. In 1961, the intellectually adventurous experimenter, Dr. John Lilly, predicted that dolphins and humans would achieve interchangeable language. Millions were intrigued by the clues he produced, myself among them. I have spent part of my waking hours since that time in the attempt to determine whether Dr. Lilly is correct or, as his critics state, guilty of sloppy thinking and promising too much.

Surely I did not *want* to find that he was promising too much. What writer would not look forward to an interview with a whale? But I wanted an honest answer, not the answer that satisfies a dream.

This book, I believe, will put you closer to knowing whether Dr. Lilly's prophecy is coming true—and closer to knowing whether various animal exotica we will inspect, such as the conversational gorillas of the Gorilla Foundation in California, are genuine "talking animals" or somehow just fake, fantasy, or misapprehension. Because the question of whether animals can master the equivalent of human speech, or even some basic fragment of it, is among the most ancient of mystery stories, I suggest that you suspend judgment while taking a trip with me through some of the newest evidence offered in the field. I don't know at what point you will feel justified in hazarding a conclusion. My own conclusion is offered, at the end of the book, but with no conviction on my part that it in any way settles the matter. How could it when principal researchers in the field and many of the fine and forceful intellects of the scientific world remain divided on the reality of interspecies communication?

A story is often told by Charr and her mother (Charr is twenty-four now) to prove that there was a time when I was more ignorant than almost anyone in the world about the communication systems of cats. This stemmed from my habit of springing up to make friendly overtures to any stranger-cat who appeared at the window, attracted by our many glowing and luscious pets.

One night an horrendously bushy old tomcat appeared at an open window, and I went over to exchange a few words with him as he sat on a ledge. The cat used gestural language, and it was not intended for me. As I knelt to be nose to nose with him, he turned around and squirted a great liquid-whish into my face—the spray intended to announce, to all female cats of the house, that he was on the prowl for love.

This was my worst personal failure in sensing the meaning, quickly enough, of a particular animal communication. Charr and her mother have never quit laughing.

I use it to assure you that we may not absolutely solve the questions surrounding animal communication in what follows here, but there will certainly be surprises.

PART ONE

TESTING THE LIMITS OF THE POSSIBLE

Man has great power of speech, but the greater part thereof is empty and deceitful. The animals have little, but that little is useful and true; and better is a small and certain thing than a great falsehood.

—Leonardo Da Vinci, *Notebook*, c. 1500

Animals are such agreeable friends—they ask no questions, they pass no criticisms.

—George Eliot, *Mr. Gilfil's Love-Story*, 1857

Anne, Anne, come quick as you can
There's a fish that talks in the frying pan.

—Walter de la Mare, 1873–1956

1

The New Breed of Animal Talk

MILLIONS HAVE COME TO BELIEVE, IN THE LAST twenty years, that whales talk to each other, and pass on great tales, just as humans did in Homeric ages when they sat about campfires and passed along the legends of the race. This popular belief has been joined by another: that gorillas can speak in sign language, when adeptly trained, and they can even become masters of a bizarre vocabulary. Gorillas have been portrayed to us as expressing a desire to squash alligators, making up phrases like "elephant devil," or telling a trainer that the trainer has the gorilla's "unattention" (as though the gorilla will only listen if it feels like it).

Does this sound like fantasy land? It comes with full scientific trappings, and videotapes and photographs are produced as evidence that there has been no mistake: gorillas talk, gorillas are voluble (in sign language), gorillas are adroit and pick up on words as useful as "no" and as specific as "stethoscope" when they are motivated to do so. The "proof" is offered, the denials ring—and great and clanging battle rages over the reality of interspecies communication in new and exotic forms.

The battle *seems* to be about "talking animals," and is, but it is also a battle about the *consciousness* of animals: their power to reason in a noninstinctive way. Using the new breed of animal talk as the greatest proof that can be offered, very determined scientists are attempting to do away with old and cherished beliefs on the limitations of animal intelligence and its confinement to the purely instinctual. With language experiments multiplying at a heady pace over the past decade, all the elements are now in place for climactic eruptions over warring scientific philosophies.

Most of us need not be so strenuous about this, though. If the clamor over apetalk and whalespeak already threatens to pit theorists against each other quite as sharply as the question of evolution did, we can nevertheless go back a step or two, look at this coolly, sort out the ideas and where they have come from—try to gain some sense of how much is fact and how much is fiction in the most recent claims for animal talk.

Whoever imagines that the truth is simply apparent will be in for a rude surprise. The new experiments can't be laughed off as mere inanities; nor have ancient and formidable objections to animal talk died away in the face of overwhelming new evidence. In the eyes of journalists and researchers reviewing all this, the wind has blown one way, then another. But there's no denying that the quest for interspecies communication has already provided surprising new developments, with porpoises vocalizing for computers and gorillas turning the pages of magazines they are asked to interpret.

When two gorillas living in a trailer-house in Woodside, California, take IQ tests, they are asked the same questions—these are put to them in sign language—as those administered to American schoolchildren. The gorillas do well, according to their trainers, although they always score lower than the average for schoolchildren. A curious event that hordes of us now accept as entirely plausible, the gorilla testing indicates just how far we have come from a comparatively recent time when primate researchers struggled in vain to teach some member of the ape family a single recognizable human word. IQ testing of the kind performed with gorillas presumes interspecies communication at a notably high level. We will have to ask ourselves: is the presumption justified?

These advanced explorations into the language and learning powers of the gorilla have counterparts in investigations into the abilities of other species. Delvers into cetacean (whale and dolphin) mentality have begun to use computers in an attempt to decode cryptic acoustical patterns which have now been recorded from hydrophonic soundings throughout the oceans of the world. Since World War II efforts have been made to reach into the dolphin's brain with electronic devices. Even the usual method of the naturalist—observing the animal in its native habitat—has been extended now. Only in our own age have there been significant attempts to swim with the great whales in order to understand what message they might have for land-based creatures.

LANGUAGE: THE BLOOD OF THE SOUL

The hope that humans can achieve conversation with the nonhuman animal dates from primordial times. There have always been legends and wonders attributed to "animals who talk," but they were quite unlike the reports of interspecies communication which exist today. In earlier centuries, they rarely came in the form of closely reasoned scientific papers. Once we had hearsay. Now we have statistical analysis.

Revolution in the interspecies quest has spurred scientists, naturalists, and imaginative interlopers into new and strange endeavors. In three swift stages, starting in the late 1950s, we have been introduced to dolphins who supposedly tried to master human words; then to chimpanzees who, using several different symbol and computer systems, were described as learning to create sentences; and finally to that pair of sign-language-using, supposedly conversational, gorillas who were able to keep up a back-and-forth discussion with their trainers.

Great ambitions stirred as the reports surfaced. Undergraduates, even their teachers, dreamed of a kind of contact with whales so profound that the whales would tell us the story of their lives in their own words. Even though this hasn't happened, it is often asserted that we're right on the verge of it—that a breakthrough is imminent and humans should ready themselves for a total intercommunication with whales.

"Language! the blood of the soul, sir, into which our thoughts run, and out of which they grow," expostulated Oliver Wendell Holmes in 1859, in *The Professor at the Breakfast Table.* Professors and students of the 1970s and 1980s frequently persuade themselves that humans are not alone in pouring "the blood of the soul" into their communications with each other.

And if gorillas, whales, and porpoises have messages of subtlety and import for each other, why not for us? *We see ourselves as the great code-breakers—let us break the code.* It is a common cry of our own civilization.

Other, and especially primitive, societies have also wanted to make this breakthrough, and yet it has eluded the inquiring minds of every age, with the possible exception of our own. We can now say that the *circumstances* for speech do seem to be present for gorillas, who, in the wilds, live in families; for whales, who at times make sounds that can travel almost limitless distance; and for porpoises, who are sometimes described as more social creatures than we ourselves. Chimpanzees are gregarious, as are many apes. Sociality in itself, though, is not a sign of complex and subtle speech. Fish travel in schools wordlessly, unsuspected of language.

Support for the belief that certain species may learn to speak with humans has been prodded along not only by formal papers but by anecdotes that have greatly popularized the idea that communication with animals is now on a level never achieved before. Sometimes the anecdotes are passed along as though they amount to hot gossip.

Could it be possible that one of those talking gorillas in the trailer-house had referred to seeing its own mother killed at the time of its capture? One of the world's foremost protectors of animals (speaking confidentially) passed this question along to me. She was very angry. On the one hand, she did not believe this, did not believe that gorillas can interpret their own capture well enough to speak about it. And her other belief was that gorillas belong in their African homeland, not taking language instruction in a trailer-house. The search for animal language sets up a complex tangle of emotions among all who deal with exotic animals, and these emotions play a role in the search for the truth beneath the issues.

We will go into the various tales—formal reports and jaunty anecdotes—to see how much they did or didn't mean about the reality of interspecies communication. A percentage of the reports are quasi-scientific or ambiguous at best. There have been others that, unless they can be explained away, do in fact point to "an age of discovery" in animal communication.

2

Compulsion and Creativity in Animal Communication

AN INBORN BIOLOGICAL OBSESSION WITH THE CONtinuation of its own life prevails in every creature at the moment of birth. This is not a choice; it is a compulsion. Language is born out of that compulsion. Once breath itself is found, the first incessant, monotonous, compulsive gestures and sounds from any creature will shout or murmur the need for food and protection.

Grizzly bears speak in a roar, birds in a tweep, but such signals as they have derive from a common need: survival. Despite the common contention that animals other than humans lack *language,* there are basic ideas connected to survival that nearly all animals can transmit in one form or another.

When prey grew scarce but tourists common, western coyotes in the national forests converted themselves into beggars. They often approach tourists with submissive sounds and gestures, looking for a handout. To those who love the coyote and its wild ways, this is a sign of the animal's degradation through human contact. It is also a sign that the coyote adapts its communication system to the nature of changing times. It can deal in signs with other species when it has to.

"Want food," "Leave me alone," "I am angry," "I submit"—the sounds or signs for such common emotions can generally be found throughout the living world. In most species they can be easily read by humans with no special training. Primitive examples of threats or "I didn't like that" are so evident in their meaning that one species may have no trouble at all in understanding another. If llamas are posted to stand guard over sheep flocks, the llamas will drive off a coyote by spitting in its direction and gesturing to indicate their intention to kick. Certain forms of communication are virtually universal in their meaning.

At the Institute for Delphinid Research in the Florida Keys (for tourists, they call it "Flipper's Sea School"), a dolphin named Little Bit gave birth to a baby before she could be segregated from the males, who generally resent newcomers. The males signaled their hostility by pressing on mother and baby, several times tossing the calf into the air. In a report on the institute's breeding program, trainer Jayne Rodriquez described how Little Bit was guided to the maternity pen, where the calf was then released. Only when her baby was safe did Little Bit turn and give "a very healthy bite to the unsuspecting handler." There was a great deal of communication here—the males were signaling their resentment of the newcomer, the trainer signaled that Little Bit must go elsewhere for protection, the mother dolphin signaled that she didn't appreciate the intervention. It is rare for dolphins to bite humans, but their ability to send messages, in many forms, is constant.

"I need your help" can be a subtler message to convey between species than "I'm going to kick you" or "I didn't like that," but many humans can verify that they have received such communication. When a vagabond cat is starving for lack of food or simply hard up for affection, a combination of "I want food," "I submit," and "I'm comfortable here" may follow in quick succession when it is discovered on the sidewalk (*it meows*), picked up (*it submits*), and taken home (*it purrs contentedly*).

When a hurricane sent its first ravaging winds through Miami, a couple in a car dumped an unwanted cat in a convenient tree and took off. Seemingly defenseless against awesome winds, the cat pealed incessantly until a woman who could stand it no longer ran out of her apartment and salvaged it from the hurricane. This developed into a long-term adoption.

Simple, understandable cries—perfectly clear between individuals and between species—can work miracles. That is the charm of language, even at its most primitive.

A DIFFERENCE BETWEEN COMPULSIVE AND CREATIVE LANGUAGE

We could call what the cat did *interspecies communication,* but that might be somewhat disturbing to scientists and investigators who give this phrase a more special meaning. The "more special meaning" is not usually defined very clearly. It really has to do with the sharp line between language that is only instinctual (that is, having to do with survival) and language that springs away into surprising paths and that is no longer a compulsive act but a creative one as well.

The linguist Mario Pei suggested, in *The Story of Language* (1949), that animals communicate through narrow systems typified by bark-and-snort communication. This signaling can take many odd forms but, Pei seemed to think, it was necessarily controlled by immediate needs and desires. Animal talk, he said, was utterly stereotyped. Humans, on the other hand, have used language to make advances in science, politics, all the arts.

"Animal cries, whether we choose to describe them as 'language' or not," Pei wrote, "are characterized by invariability and monotony. Dogs have been barking, cats meowing, lions roaring, and donkeys braying in the same fashion since time immemorial. The ancient Greek comic poets indicated a sheep's cry by Greek letters having the value of 'beh'; in modern Greek, those letters have changed their value to 'vee.' The sheep's cry has not changed in two thousand years but the Greek language has. Human language, in contrast with animal cries, displays infinite variability, both in time and in space. Activity and change may be described as the essence of all living language."

Those who would deny that humans alone have "living language" are now armed—as they weren't in 1949—with a spectacular but solitary piece of information that seems to refute the assertion that animal communication never shows any change or evolution whatsoever. Refutation came in the form of underwater recordings, which brought us the changing sounds, year by year, of the humpback whale.

The humpback, as we will examine in more detail later, returns each year to such favored spots as its Hawaiian birth grounds with a different song. This year's song is different from last year's.

Last year's was different from that of the previous year's. Next year's will be different still. At first, it was thought that it might simply be a matter of different whales—one tribe had a particular song and another tribe a different one, and so on. The recordings changed minds. Humpbacks actually were showing "variability in time and space," at least in this one aspect of their being. A case could be made that the whales' changing refrain does not constitute "infinite variability" and that the seasonal modulations are in fact quite limited. They do, nevertheless, clearly show evolution in an animal sound pattern.

What was the difference, anyway? And why should such changing refrains matter? They matter because they raise the possibility that within the animal kingdom there may be other creatures besides humans who are not merely compulsive in their communications but creative as well. Darwin called language "half instinct, half art." If what animals say and how they say it is half instinct, or even more than half, another part *might be* art—a changing creation by the individual or the pack. Are we absolutely sure that wolves howl today as they howled in ancient times? Can we be certain that no wolf innovator has ever come and turned the howling in a new direction? Can we truly outlaw the possibility of any aspect of "living language" from the cries of all animals everywhere?

A new breed of interspecies communicators would say that we cannot, that signs are now accumulating that we have seriously underestimated the variability in animal communication. They wouldn't suggest this of all animals, but only of a few species—those same species we have usually considered the brainier ones. Gorillas, chimpanzees, and whales are leading candidates and the chief subjects for experiment and inquiry.

HORSES AND THE TRANSFER OF MEANING

Older investigations into interspecies communication concentrated on horses and dogs, and there are still defenders of the view, as there were when Pei was writing *The Story of Language*, that it is possible to find creativity in the way a dog or horse will reason—and communicate its reasoning.

The English horse breeder Henry Blake concluded that he could understand most of what horses were saying to each other because their signs were so specific. "We can interpret what our horses are saying to us and to each other, and we can make our horses understand what we are saying to them," he claimed.

If the latter were not true—that humans can convey their meaning to horses—it would not be possible to ride, or at least not possible for anyone who did not have the skills of rodeo performers. Body language between rider and horse is very intense. Even if the rider doesn't realize that he may be communicating fear and indecision, the horse senses this with considerable accuracy. Not wanting to be in the hands of a frightened or inexperienced person, certain horses will summarily dump such a rider. There hasn't been a failure to communicate; in fact, the rider has communicated his inability all too well.

In his book *Talking With Horses*, Henry Blake took horse language one step further by enumerating a "dictionary of horse language" consisting of forty-seven familiar signs that humans, if they know them, can read almost as easily as horses can.

The human who can read horse language need not be taken by surprise when the horse bucks, because, according to Blake, it sends preliminary signals to convey its intentions. To warn "I will buck," says Blake, the horse "points his ears sideways and arches his back and makes as if to buck." The similar but more abstract message "Go back" can be conveyed when a horse swings its head, which other horses easily recognize as an indication that they are not to pass.

"Twitching the skin and possibly waving a leg, stamping, or squealing," according to Blake, often just means "that tickles." Humans would have to study horses quite closely in order to know this.

The submission sign of the foal, which derives directly from the instinct for survival (large unfriendly horses could kill a foal), is made, says Blake, by "holding head and neck out straight, sometimes holding the nose up slightly and moving the mouth as if sucking. When [the foal] does this it is most unlikely that another horse will hurt him."

Blake had a way of translating horse signs into phrases that imply a bit more than "I submit." That mollifying gesture from the colt, for instance, is described as the proclamation "I am only small." Blake's book appeared in 1976, at a time when doubts and ambitions on the topic of interspecies communication were greatly multiplying. Where the scientist might require strenuous proofs that any given set of gestures could be added up to the single emotion "fear," Blake was entirely offhand about it, asserting that anyone can check his observations by going among some horses for a trial.

The charm of his book lies in the pleasantly colloquial, sometimes specifically British, ring that he ascribes to common equine expressions. When he goes out to greet his horses, he says, these are some of the common communications he can note: "Come and drink," "Come and fight," "I am king," "I am tired," "I suppose I will have to," "I cannot," "I am enjoying this," "Gangway," "Give me some more," "I love you," "Let's get the hell out of here," and "Where's my bloody breakfast?"

Columbia University psychologist Herbert Terrace, one of the researchers whose own work with interspecies communication we will examine later, says that horse signs fall short of being language because horses have but one way of describing an emotion ("A horse seems able to make one and only one sound when it is frightened"). Are horses, then, never creative or varied in their language? Blake believes they are. For Blake, many "creative" communications with horses take place on a psychic level. But even at the sign level—and if his discussion can be taken as definitive—horses exhibit genuine spurts of invention in communication.

Every horse, says Henry Blake, signals "It is good to be free" in its own individual way. Compulsion yields to creativity.

TALES OF PETS: WONDERFUL BUT UNCHECKABLE

While primate behaviorists and dolphin investigators were trying to devise experiments that would prove elaborate forms of interspecies communication beyond any reasonable doubt, pet owners and animal handlers of every kind came in with tales far more remarkable than anything that laboratory-based interspecies communicators were able to establish. This has assuredly been happening for as long as humans have taken animals into their homes.

J. Allen Boone, who had no connection with the mysteries of animal communication until he was given an entirely chance assignment to be the night-and-day companion to a movie-star dog named Strongheart, has demonstrated in a book called *Kinship with All Life* (1954) just how far a confirmed belief in the communicatory power of animals can take the convert.

Boone seems fairly alone in his contention that humans can achieve complex communications with flies. He claimed a fly he called Freddie would land upon his skin only on invitation. Boone said he had a game he played with his personable fly that involved giving Freddie directions to soar for "landing-field areas" on one of Boone's palms. This smacked of tall tales, but Boone presented his view with the earnest mysticism of an Eastern religionist describing the inner workings of the universe. Boone insisted that "various four-legged, six-legged, and no-legged fellows shared priceless wisdom with me." The main piece of wisdom appeared to be that humans would do well to respect the intelligence of nonhuman species.

Boone's description of the first night he ever spent with Strongheart, a champion German shepherd imported from Germany to become the star of such films as *The Silent Call* and *White Fang*, is a trifle more elaborate than tales from the usual pet raiser, but it is different only in degree. Boone decided that he and Strongheart would spend the night in the same bed in order to speed the process by which they gained mutual understanding, but Strongheart proved an uneasy bed companion. The shepherd was repeatedly off to investigate sounds in the night. Each time it returned from these forays, the dog crawled into bed with his tail at the head of the bed, and he resisted every effort to persuade him to put his tail the other way around.

When Boone became firmer in his commands, Strongheart

> suddenly moved forward, closed his jaws on one of my pajama sleeves, and tugged me gently but firmly to the foot of the bed. A short distance away were some rather insecure old French windows covered by long curtains. Getting an end of one of these in his teeth, Strongheart pulled it back, held it there for a few seconds, and then let it drop back into position again. Then he began barking, swinging his

head rhythmically back and forth between the French windows and me.

Had he spoken in perfect English he could not have told me more clearly what he wanted me to know: that whenever he lay down to relax either temporarily or for his night sleep, he always wanted to have the front end of himself—his eyes, nose, ears, and jaws—aimed in the direction of possible danger. If trouble did enter in human form, he could go into action without having to lose time by turning around. And those old French windows were certainly a possible trouble entrance.

Why not take a tale as lively as this as a perfectly acceptable proof that, whatever the limitations of other animals, dogs are perfectly capable of ingenious, creative, highly individualized acts of communication? Those who experience them usually do accept such incidents as the proof of something important, although they may not give it a label as imposing as *interspecies communication.*

But most such tales are highly subjective and usually unrepeatable. Humans can use flights of language, spoken or written, to describe how to protect a bedroom. The dog's "explanation" is a story remembered. Few of us wire our bedrooms for sound and set up movie cameras to record otherwise elusive evidence of animal communication. (Those who do this at all usually have something other than a dog's behavior in mind.) Such tales remain anecdotal and cannot alter accepted scientific belief without repeated demonstration.

THE BIRTH OF AN INTERSPECIES MESSAGE

The belief in at least some form of interspecies communication has spread at a greatly accelerated pace in recent years, even among the scientific observers with high standards for the admission of evidence. This turn of opinion has come about in part from the experiences in training dolphins. Suddenly there were "leaps in learning" that proved to nearly all concerned that these particular marine mammals can grasp human communications at an advanced level.

In order to train those small cetaceans we usually call dolphins or porpoises, humans must communicate what they want, but they have no language for this. If the trainer is experienced and inventive, the difficulty can be overcome. The *language* is invented between trainer and subject out of the materials at hand. Although some signals have become standardized, each trainer tends to have a quite personalized method, a system for establishing brain-to-brain contact. At the beginning, an animal brought in from the ocean can be entirely innocent of the meaning of human sounds and gestures, so the trainer must bridge this gap. And the trainer is very much alone—alone with the whale and with the problem of inventing language. What the human understands is of little value. It is necessary to discover what the animal understands.

In the fall of 1978, two melonhead whales were captured in the ocean and brought to Sea Life Park, an ocean exhibition center on the island of Oahu, Hawaii, for training. The park has become an excellent testing ground for some unusual ideas on whale capabilities. This is what happened when the melonheads, called Lahaole, meaning "first of a kind," and Kahuna ("sea witch"), were turned over to Ingrid Kang. Born in Sweden and trained on a scholarship at the University of Washington in zoology and animal behavior, she became chief trainer and curator of marine mammals at Sea Life Park. She had worked there since the mid-1960s. Star performers like the Pacific bottle-nosed dolphin yielded quickly to her methods. The park is apt to have twenty dolphins at any given time. Most are not unfathomable creatures.

The melonheads were.

Although she had become one of the most adroit instructors of dolphin in the world, Ingrid Kang ran into immediate roadblocks when she tried to "solve the communication problem" with these low-in-the-water specimens from a species not dealt with before. A Pacific bottle-nose had been captured at the same time, and it was soon up in the air performing. How to cross that barrier with the melonheads to let them know what their human captors expected of them? They were different from the bottle-nose, very different from humans, and even different from each other.

In order for one creature to contact another, there must be something they have in common—something so well understood that it can become the symbol of the meaning to be exchanged. Bottle-nose dolphins, the creatures who seem to learn fastest, come into communication speedily in response to food offerings—tasty herring and smelt bring them to the surface. Then, as a particular dolphin displays its natural acrobatics at the top of the water, the trainer has rewards and praise for the kind of acrobatics that will develop into spectacular stunts. This is called positive reinforcement. If the dolphin is lazy or misbehaves, the trainer may turn away as though uninterested (negative reinforcement). Like a child in its growing-learning phase, a dolphin tends to fight for attention, and skilled trainers use this tendency to bring out the behavior they like. The humans are communicating, but it is debatable—it depends on the precise actions and their motivation—whether dolphins are behaving compulsively or creatively from their own end of the communication pattern.

Profferings of fish failed with the melonheads. They stayed underwater. None of the signals that had worked with other marine mammals had meaning for the melonheads. In the course of trying to figure out what the melonheads might understand, Ingrid Kang finally made a discovery that enabled her to bring them to the surface, doing some sporty action. But what do humans and whales have in common that can become the starting point for understanding if even the ancient "I'll feed you" ploy has failed?

What Ingrid Kang found was: rain. Humans and whales, creatures of two universes, have the rain in common.

Only when it rained did the melonheads come to the surface and demonstrate a variety of natural acrobatics where humans could observe them. This would not have been a useful discovery for the trainer unless she also found a way to say, "Act like you do when it's raining." Her first step in communicating this message was to bring a water hose to the pool. By simulating rain with the hose, the melonheads could be brought to the surface, where they then ran through a repertoire of natural behaviors—natural for them in the rain, that is.

Soon the melonheads understood that if the hose was used it meant they were to go into action.

The amazing part of this discovery—we might say *the language part*—was the way in which Ingrid Kang was able to telescope her original finding. She tried putting the hose aside ("I faded out the stimulus," she reported in a paper written for other dolphin trainers) and just splashing the water with her hand. Later she didn't splash at all but merely waved her hand over the surface of the water. Now, from a wave of the hand, the melonheads were producing the same behavior that had once happened only in the rain.

In re-creating the sequence of interaction between Ingrid Kang and the melonheads, we have witnessed the birth of language. At this point the language is quite simple—it involves only one understandable sign. But that one sign condenses a great deal of meaning and actually constitutes a language link between species. A hand wave has become the symbol for a thought that, if expressed between humans, would require a complete English sentence something like: "Do the stunts for me right now that you did when it was raining."

HOW THE DOLPHIN SPOKE TO THE TRAINERS

The search for some kind of spoken language between humans and dolphins has sometimes obscured the extent to which a developed *gestural language* has grown up between particular trainers and particular dolphins.

Bottle-nosed dolphins learn to pick up what the trainer intends from very subtle signs. If the trainer wishes, spectators may never guess how instructions to the dolphins are being communicated. When she was asked what communication methods work best of all with her most talented dolphin confederates, Ingrid Kang told me, "Eye motion."

I've used the word *confederates* here because Ingrid Kang, like most dolphin trainers, describes the training process as a reciprocal relationship rather than the *I-command, they-obey* stance usually found between animal trainers and their subjects.

She feels that certain mammals in her charge accept her as the leader, but others do not. She communicates with them "at a fifty-fifty level." By exchanging those signs that both parties understand, agreements are reached, which is no different from the way training proceeds between parent and child. The nonnegotiable parent who is all-boss soon can be in deep trouble in trying to determine the child's behavior, for the child has a will of its own and has to be admitted into the bargaining process.

But dolphins don't speak human language and humans don't speak dolphin language, although many believe that such a language exists. How can the creature in the tank break through to the human and explain its own "ideas" to the creature (the trainer) who has first appeared in the position of instructor? The dolphin is in somewhat the same position as Ingrid when she was searching around and then found rain as something the other creature would understand.

Dolphins know, by the behavior of trainers and crowds, that something humans understand is *tricks.* Dolphin trainers generally agree that when dolphins speak to humans, they do so by offering tricks—not just those the humans have in mind but new ones the dolphins believe might be of interest. As extraordinary as this may sound, dolphin trainers have claimed that there is a point at which dolphins take over in the training sessions and begin to audition new tricks and new ideas the trainer may not have been asking for. When J. Allen Boone claimed that he had been trained by the dog Strongheart rather than the other way around, it may have sounded unduly romantic and sentimental, but dolphin trainers very commonly remark, "The dolphin trained me." They do not mean that all the training came from the dolphins, but that some of it did, and their comment is intended seriously.

An auditioning dolphin trying to please its trainer with something special and new can go into a new flip in the middle of an old trick as though to say, "What do you think of this? I just thought of it." The Swedish authors Karl-Erik Fichtelius and Sverre Sjölander concluded (in *Smarter Than Man?*) that dolphins and their relatives may be the only creatures besides humans themselves who are able to grasp the principle of "novelty."

Even if communication between dolphins and humans has remained at the demonstrating, gestural level, a strong argument can be made that communication between these two species has proven to be artful as well as instinctive, and raises the prospect that communication can go much further.

When the contact between the trainer and the dolphin has reached a certain stage, there is enough experience to almost guarantee particular results. When Ingrid Kang first persuades a bottle-nosed dolphin to put its head out of the water and keep it there, she knows that in just about a year the communication between the two of them will be so perfected that the dolphin will be leaping eighteen feet into the air at even the tiniest of human signals. But those eighteen-foot leaps are not something the dolphin would make a regular practice of, or have any need for, in the ocean. They are the dolphin's answer to the human communication of a very human desire: *Let's wow the crowd.*

THE COMMUNICATORS' MAJOR CLAIMS

Although aquarium spectators usually don't know it, the ocean-born dolphins whose whistlings, clickings, pulsings, and squawkings they hear did not use such sounds during their spurts out of the waves when they were still wild creatures. Those above-the-water sounds are a response to the human desire to hear them talking. And because the sounds are so varied and the dolphins' vocal system capable of so many types of sounds, humans have wondered if the dolphins could understand a human desire to have vocal interchanges with them just as they grasped other ideas that came from the human race.

A dolphin named Lizzie was the first creature from her species to be credited by her trainer, neurophysiologist John Lilly, with making what he called "humanoid sounds." Starting in 1957, a set of communication experiments was carried out with Lizzie and a tank mate named Baby in dolphin tanks located at St. Thomas in the Virgin Islands. Dr. Lilly's methods of prompting dolphins to imitate human speech included artificially stimulating regions of the brain by implanting electrodes. The sounds Lizzie and her successors produced were very different from those typical of dolphins. Sounds seemingly produced through the dolphins' blowholes included what Dr. Lilly described as a Donald Duck–like quack.

Dr. Lilly's career and experiments, which we will examine further in the next chapter, have provided a principal incitement to explore the possibilities of dolphin talk. A startling observation was made during intense study of recordings made of Lizzie and Baby over a long period: Dr. Lilly found a single dolphin interchange in which Lizzie and Baby seemed to be using humanoid sounds to each other underwater. The implication could be that the isolated dolphins were taking a bit of the human communication system into their own repertoire, although Dr. Lilly, like others commenting on his work, leaves this as unproved.

On the night before she died, Lizzie heard Dr. Lilly call, "It's six o'clock." Lizzie seemed to answer him. What she said, Dr. Lilly claimed in a 1962 lecture, "sounded suspiciously like, 'It's six o'clock.' "

Dr. Lilly later moved his operations to Miami, where a young male dolphin named Elvar reputedly produced sounds like "ir," "squir," and "irt." Lilly was trying to teach him the word *squirt*. Elvar, as Lilly saw it, could be heard using human sounds as though attempting to analyze their content. Although these "humanoid sound experiments" were given up after a time as not leading far enough, and Dr. Lilly came to regret the electronic meddling into the dolphins' brains, the experimenter made a remarkable inference along the way. He concluded that the dolphin Elvar interchanged humanoid sounds with sequences of dolphin sounds "as if to translate for us." It was Dr. Lilly who asserted that dolphins have a language and christened it Delphinese. He did not claim to understand it; he simply said that it was there and that humans (or computers) might decode it.

Dr. Lilly's emphasis has shifted from trying to teach dolphins human sounds to trying to fathom dolphin communication, on the one hand, and dolphin ethic and dolphin character, on the other. The claim of Dr. Lilly's that has divided all dolphin experimenters of the past two decades is his 1961 proclamation (in *Man and Dolphin*) that humans and another species will learn to communicate in a common tongue. Even though he was speaking of the possibility of communicating with extraterrestrials, he clearly felt that the unknown creatures we are most likely to speak with first are the greater and smaller whales.

The notion that dolphins and humans could learn to compare notes on their two different worlds has been vastly exciting to many who follow the work of animal naturalists and innovative behaviorists. The growing popularity of ocean parks has made it easy to see what Dr. Lilly was talking about when he described the virtuosity, even "the ethic," of the dolphin. Lithe young aquanauts, male and female, scrambled onto the backs of dolphins and stuck their heads into the mouths of killer whales, and yet these animals behaved with dependable gentleness. Since the killer whale is one of the great marauders of the sea and the dolphin can bull the shark to make a snappy exit, their almost unfailing benevolence toward humans demonstrated either a great sensibility or the quietly intelligent deduction that biting or butting the host could have undesirable repercussions. When a dolphin wanted humans to get out of the tank, it usually bunted them with a series of proddings calculated not to harm—very firm but very gentle. Dr. Lilly began to say that this particular set of mammals might well prove more intelligent and better-natured than humans, greater creatures than ourselves.

Clever Chimpanzees, Slangy Gorillas

Primate experimenters who have pursued innovative communication experiments, starting in the late 1960s, don't present even their keenest animal students as superior to humans in intelligence, or as rivals to them. They do describe extraordinary results, far beyond anything believed possible only a few years ago. Sentence-making chimpanzees and conversing gorillas jumped from folklore to seeming fact. American Sign Language (ASL, or Ameslan) provided the first breakthrough when a chimpanzee at the University of Nevada was introduced to the same hand signals used by the deaf. When this seemed successful—the experimenters reported extensive vocabularies, a creative use of words, and occasional complete thoughts, expressed in sentences, as coming from the chimpanzees—additional symbol systems were tried. All the systems depended upon giving the apes substitutes—in geometric, abstract, and computerized forms—that corresponded to words or expressions found in human language.

At the Yerkes Regional Primate Research Center in Atlanta, computer-using chimpanzees whose sentence making was monitored by methods intended to guarantee authenticity appeared to pick up a rudimentary knowledge of language with considerable agility. A chimpanzee named Lana developed a reputation as possibly the most gifted of the "talking chimps." Lana knew the word "tickle" and often expressed requests for tickling to her trainers. She also knew that the word "machine" referred to her computer. One lonely night she connected the two in the soulful plaint: "Please, machine, tickle Lana." It was immediately famous. There was a great desire to believe.

Critics soon questioned whether these apes were as clever as they seemed or were essentially imitative and just churning out rote sentences rather than new and creative thoughts in the way a human expresses them. This attack on the significance of the results in a whole group of chimpanzee experiments has been gathering steam for several years. Some of the researchers did not wilt under criticism, and it has even been argued that the first phase of the chimpanzee experiments is successful and that the real question now is whether a first generation of sign-using, symbol-using, or computer-based chimpanzees will pass on their skills in syntax to their offspring.

The claims grow larger. The stakes increase. Prestige is on the line. Tied to the prestige is the life blood of research: grant money. Applause or ridicule from one's fellow scientists is no longer a matter simply of one's *amour propre.* It is a matter of the survival of one's work, for those colleagues and peers sit on the grant-review bodies that dispense the largesse on which scientific work today is largely dependent.

Breakthroughs in these experiments are followed with breathless interest. Fame or notoriety can await the most extraordinary claimants. Reports of successes (real or imaginary) in interspecies communication have been on an ascending scale since Dr. Lilly's 1961 prophecy. During the 1970s, at Woodside, California, these culminated in the most decisive claims ever made, in a scientific context, for an extraordinary forward leap in animal communication.

Two gorillas trained in American Sign Language appeared to go much farther than dolphins or chimpanzees have ever gone.

The trainers depicted the gorillas as slangy, deceitful, and capable of making up lively epithets. One of the gorillas supposedly called the photographer who dogs their footsteps a "devil toilet." Francine Patterson, the Stanford-trained developmental psychologist who originated the gorilla training, described gorillas as bursting with unexpected thoughts. Michael, the young male, reported a dream, she said, in which coyotes were attacking the Gorilla Foundation (his trailer-house home). She believes he tried to tell them (gorilla signs have to be interpreted) that if anything like that happens, he is ready, and will fight off the coyotes to save the humans.

Now this was on a different scale than all previous reports on the kinds of exchanges that are possible between animals and humans. Gorillas reporting their dreams, under scientific auspices or otherwise, are new in the world—new since the late 1970s, when this kind of gorilla talk was first reported.

STUBBORN SECRETS

Translating dolphin. Computer-chattering chimpanzees. Gorilla security guards ready to save their trainers if their home is attacked.

Surely all this is enthralling. But is it real? These scenes are so much a part of what we have been led to expect of the twentieth century that we hate to put them aside as a modern form of myth.

Claims of progress come in profusion now with the dropping of the morning newspaper upon our doorstep. The talking gorillas appear on the TV screen, a bit mugwumpish in the way they pause and grope before answering a question but seeming to pass tests with flying colors. Divers equipped with strange musical instruments go into the ocean, hoping to prove to us that whales are mighty talkers and seeking some special mode of communication—is it music?—that whales will understand and respond to.

Whatever the inquirer may think of this quest, it is a search of considerable fascination—like poking into caves where scratchings on the wall hint of lands and times that hold their secrets stubbornly, and for which we have no Rosetta stone to unlock their mysteries as we unlocked the mysteries of vanished pharaohs.

For those who take it up in person, it is a perplexing search, but an inspiring one. Investigators take serious risks when they advance into the homeland of the mountain gorilla or invade the sea paths of the greatest of the great whales. Nor is there anything placid about dealing with a gorilla in a trailer-house.

What evidence can the jungle explorer and laboratory explorer hope to produce?

There are four distinct possibilities in "animal conversation" that have significance for interspecies communication:

1. A human speaks to an animal—and the animal understands.
2. An animal speaks to a human—and the human understands.
3. The animal and human speak to each other—and they both understand.
4. Two animals speak to each other—and a human is able to translate.

Not all of these possibilities hold equal weight, and they do not necessarily inspire the same degree of wonderment or dispute. Even the fourth kind of encounter holds strong affirmation for the prospects in interspecies communication because once we establish the existence of genuine conversation rather than merely compulsive, instinctual signals, our ability to join the conversation might depend, as Dr. Lilly has claimed, simply on dexterity and further decoding.

A CONTROVERSY OF DARWINIAN DIMENSIONS

Casual experiences, accidental experiences, perhaps overestimated experiences, very personal and individualized experinences—we have already looked at some experiences of horse trainer, dog trainer, whale trainer—have taken twentieth-century investigators into a new era of suspended disbelief.

Every clue to the reasonableness of a faith in interspecies communication is being zealously inspected and debated. There is a great determination to try and take the "ifs" out of assertions in this field.

Those who are most in doubt as to the reality of the interspecies theorem don't express much skepticism until the phrase *interspecies communication* is expanded to mean: "Animals talk in the sense that humans do" or "Animals have an extensive and flexible vocabulary, full of symbolism" or "Animals can learn language if humans will make the effort to teach it" or "Animals are capable of conducting a conversation with humans, but humans will have to master the animal language systems."

We have already noted that a basic ability in communicating with humans can mean life or death for a cat marooned in a city or a coyote starving for lack of natural prey. Even at the compulsive, instinctual level, the flexibility of animal communication can be demonstrated. But coyotes do not go into politics, and cats do not become teachers of French. Even dolphins have not become direct spokesmen, on their own behalf, against the letting of permits that allow tuna fishermen to kill them.

The feats that come from language alone—and language of "infinite variability," expressed in terms that humans can understand—are still presumed to belong to the human race alone. Is that period ending? Will we ever truly pass to an age in which whales, gorillas, or other candidates for advanced degrees in communication begin to enter realms where more than a modicum of dexterity and creativeness is required?

That "gorilla breakthrough" into conversation, if real, could tell us something quite new about the dimensions of animal consciousness. It could even speak to us about human consciousness. Many of the investigators believe that the study of the higher primates may lead to a clearer evaluation of that primate, homo sapiens, which takes itself to be the highest of all creatures. The assertion of some whale investigators that whales, once spoken with, will prove to have a greater and more benevolent intelligence than we ourselves do rings mockingly on the ears.

Is there anything to it, or is it simply romanticism breaking into fresh excesses? Are any of the interspecies "breakthroughs" truly fact rather than fiction or wish fulfillment? Could they really be the work of fabulists, creating great visions of "talking animals" who, in the long run, will be found not to talk? For reasons connected both to the future of life on this planet and to some of the missing chinks in the theory of evolution, this is a question of paramount importance as well as paramount interest. The quarrel over interspecies communication has threatened to become an explosion as great as the explosion over Darwinism itself. And it has much the same roots.

Whether genuine interspecies breakthroughs have occurred is not as easy to establish as it might seem, because those we would like to call upon as expert witnesses in the matter are largely in disagreement with one another. We *should* be able to move, though, like a safari party going through jungle underbrush, and eventually find our way to some clearing beyond the disagreements. It is time, and past time, to do this.

Let us enter the underbrush and see where we come out.

3

Dr. Lilly and the Dark Gulf of the Unknown

DR. LILLY IS IN PURSUIT OF ONE OF THE GREAT scientific mysteries of the ages. Others have done this before him. Many have even won out, shattering ancient puzzles in the process.

While tracking John Lilly's search for communication with other species, I kept thinking about James Watson, an untried scientist who fell into the search for the structure of the secret life-determining molecule that contains the genetic code for what each of us becomes. Watson found himself in a band of celebrated scientists who probed, disputed, gossiped, fought—that is exactly how he himself portrayed them—and made their way toward the discovery of the double helix and a Nobel Prize for their efforts.

The double helix could not hide. After all those billions of years, it had too bright a scientific posse on its tail. Bright minds, applied to a purely physical puzzle, trapped it. Only those with an acutely informed appreciation of chemical process and molecular structure can gain any true sense of what the double helix is. But we know it is there, anyway—a Great Reality, so to speak—and we can feel a thrill that the search for the structure of DNA succeeded.

The quest for apetalk and whalespeak has a different air. Anyone can readily visualize talking apes and sermonizing whales (not to mention frog entertainers and glamorous porcine beauties who write best-selling guidebooks to modern living). The phenomenon of interspecies communication is not, like the double helix, a self-evident structure of the physical world that can be pinned down and labeled. Our pursuit can be waylaid by our prejudices. A very large posse, larger than the one that chased the double helix, is presently engaged in a battle to demonstrate that interspecies communication even exists.

Where, exactly, does the search lie? Somewhere between the search for the double helix and the search for the Holy Grail. And it tends to resemble both. Somehow, in the course of the quest, pure science and pure mysticism have together become its parents.

We could say that the search for the double helix began in that first moment when primitive man wondered why all of his kind were a bit the same and yet all so very different. But we also know that the search only went into its final, productive phase when Watson, the erratic but ingenious postgraduate, arrived in England, his mind dazzled with the notion that the truth was probably ready to be discovered.

In just the same way, the search for apetalk and whalespeak has been always with us but what may be its final, productive phase dates to some moment very near at hand. Since science likes precision in these matters, I will date it to the moment in 1960 when Dr. John Lilly was completing the manuscript of *Man and Dolphin*.

At times, like Watson, Lilly was bold. He dared to contradict other scientists who knew the nature of his search: scientists who were more celebrated than himself, more skeptical than himself, considerably less obsessed than himself. At times, too, he realized the limb he was on and drew back away, introducing some cautious proviso to prove that he was the responsible researcher to the very bone.

"I hope I have made it clear," Dr. Lilly writes in *Man and Dolphin*, "that I do not intend to defend a single-minded thesis or accept the provisional working hypothesis of a whale language as anything more than a temporary aid to research design. I have always qualified my viewpoint as an exploratory research position: until the experimental data are collected and fully worked over, an open mind is essential and the number of hypothetical positions remain (at least meta-theoretically) infinite."

Yes, but meta-theoretically speaking, Dr. Lilly was in love with the idea that whalespeak was out there and knocking, ready to disclose itself in wonderful fullness at any moment. And he was positioned as the most likely discoverer.

Who is Dr. Lilly, and what has he been up to? By his own description, he is an Iso-Human from the eighth level of intelligence. If we follow his own career for a bit, we will discover what he could possibly mean by this, who the Iso-Humans are, and we will also find that Dr. Lilly has been the generating force who took the search for apetalk and whalespeak into its current phase.

If he resembles any creature in those sea worlds he investigates, it is not the porpoise with its laughing face, but more likely the swordfish. Painfully thin, John Lilly can seem at times to be all beak—a probing intelligence, angular in thought as he is angular in body. When he speaks, often in a run of quips and half-embittered drolleries, again there is that sense of a swordfish thrusting. Lilly's jaw, taken by itself, is an arrowhead. It is sharper than all the other sharpness to be found in his bony frame.

He is an abrupt, brilliant, sinewy presence, and that presence can be felt in the background of nearly all recent discussions on the ultimate reality of interspecies communication. Certain basic speech patterns of Dr. Lilly have begun to show up in the habits of others who wish to see the whale and dolphin as he does. I could feel it while interviewing for this book. The Lilly-like phrases sound like this: "I think the dolphin is amused by us," "I get the feeling the whale would know more about that than we do," and "Let's hope the animal is dumb enough to buy what we're saying, should we?"

At a long weekend study session at the Hyatt Regency in San Francisco, a building whose interiors, in their striking diagonals, somewhat resemble the picture I have of the mind of John Lilly, I watched him deal with dozens of his own faithful. At last he was asked why humans should bother to seek a communication with creatures beyond ourselves. He gave a Lillyesque answer. He said that humans are lonely in the universe, they need someone new to talk to. That loneliness, he seemed to suggest, could end.

Though romantics in the interspecies movement easily picked up the speech habits of Dr. Lilly, they usually lacked that tigerish ability to set up powerful theoretical questions in a new form and conceive practical experiments that show some possibility of providing answers. He has the style, and the innate confidence, of the great innovator who discovers his own purposes out of the confusion and bafflement of those who fear to leave the safety of the known. The swordfish in him often comes out—he asks questions, sharp and skeptical, which throw his own followers off balance. How do we know that whale language, when discovered, will be in a form that even the cleverest humans can follow? What is your picture of language? Do you realize it could take forms you might mistake for something else entirely?

In *The Mind of the Dolphin* (1967), where he feels a need to explain the numerous possibilities, Dr. Lilly notes those areas where whale and dolphin communication with humans might be accomplished, suggesting we not be narrow-minded on the form an interspecies communication can take. While he cannot describe transmission systems that humans themselves don't have and therefore can't understand, he suggests that we do have a start on the transfer of meaning by various systems—not all of them usually understood as communication mechanisms—which humans are familiar with.

"Among the modes which do exist," Dr. Lilly wrote, "are conventional speech; gestural, nonvocal transmission and nonhearing reception; some kinds of music; writing and reading; mathematics; dancing; lovemaking. Many of these modes are not usually contained within a definition of communication. The new science of nonvocal communication between humans includes those activities." Lilly does not rule out the telepathic, either, but suggests that we leave questions of thought transference "to future research."

How did he come to be the original pacesetter in a newsprung advance on the secrets of whalespeak? By training, by experimental record, by natural disposition. If the search for apetalk and whalespeak has been, as I suggest, a twinning of practical and mystical concepts, that was right down Lilly's alley—rare among scientists whose great talent is for practical investigation, he has been unable to let the mystical alone. At first this questing after secret structures in the universe took a mean form: as brain investigator, he began planting electrodes into the brains of cetaceans in order to artificially control behavior. Later, he had a great revulsion from this kind of experimentation and stopped it. His career and personality are full of paradoxes.

Dr. Lilly began in neurophysiology as a graduate of the California Institute of Technology and the University of Pennsylvania Medical School. His fascination with marine mammals began when, as a United States public health officer, he undertook an assignment to map the brain of a dead whale. Dr. Lilly's own brain was provoked, first, by the impressive size of the cetacean brain—the larger whales have seventeen or eighteen pounds of brain, as compared to the average human's three-pounder—and, second, by its striking structural similarity to the human brain. Later neurophysiological studies revealed the significant fact that the dolphin brain, like the human brain, possesses large cortical areas known as "silent" areas, which in humans may be responsible for many of the creative functions of intelligence such as long-range planning. In his essay "The Dolphin Revisited," published in *The Dyadic Cyclone*, Lilly offers this explanation: "Presumably these are the areas of our brain in which we do our major central processing (computations) as humans. That which we value most as humans . . . is in these silent areas. They are the association areas for speech, vision, hearing, and motor integrations and for relating these to all other activities of our bodies."

The existence and size of the silent areas is one of the primary features that distinguishes the human brain from the brains of other nonhuman primates. Lilly's observations about the cetacean brain led to his hypothesis that marine mammals possess great intelligence—possibly equal to or greater than human intelligence.

In addition to intelligence, Dr. Lilly's brain studies led him to expect something deeper, which he calls intellect. Probably there is no clear way to describe any nonoverlappingness between intelligence and intellect, but a hint of what Lilly is thinking seems to lie in the difference between the idea of being deeply thoughtful, profound in the deeper areas of existence.

The fact that Dr. Lilly considers such distinctions—not a matter certain to be high on the agenda of the average behavioral scientist—may indicate why he happened to take the road that he did. In the 1950s, realizing that a quest as unusual as his would not find ready support in institutional monies, he sold all his properties to set up a dolphin lab in the Virgin Islands and founded what he called the Communication Research Institute for Biomedical Studies. He wanted to talk to dolphins and thought it might happen. In the lab at St. Thomas, and later in Miami and California he pursued a series of innovative experiments—far different from those anyone else had attempted with dolphins—that were highly illuminating even when they failed.

At the time Dr. Lilly began his study of the bottle-nosed dolphin in 1955, he viewed himself as the detached scientific observer. By 1957 the observer had made the startling observation that the dolphins he kept in his laboratory tanks for research purposes had the capability of mimicking the human voice. Tape recordings of dolphin sounds, played at reduced speed, slowed down the high-speed, high-pitched dolphin vocalizations so that they could be recognized as humanoid imitations.

Lilly and his colleagues at the Communication Research Institute also discovered that dolphins brought into captivity seemed to initiate contact with humans. In the ocean, dolphins emit sounds only in the water, but after several weeks in captivity, the dolphins gradually began to emit their whistles, clicks, and chirps in air, often seemingly aiming at a particular individual. Writing in *The Proceedings of the American Philosophical Society* in 1962, Lilly noted: "After several weeks of such noises one begins to notice a changing pattern of the airborne sounds to more complex sounds involving longer emissions, greater richness of selection of frequencies and harmonics. In our experience such changes occur if and only if people have been talking to the animals directly and very loudly individually. Slowly but surely these sounds become more and more like those of human speech."

Dr. Lilly, a state-of-the-art scientific practitioner, was able to encourage the production of these humanoid sounds through the use of electronic brain implants. His description of this technology, in *Man and Dolphin*, rings with a tolerance for brain meddling that he would later abandon. "In the less well-controlled training of the past we depended solely on a fish-food reward," Lilly stated. "With the newer method we can push a button and cause a brief controlled period of a rewarding pleasure to occur to the particular animal. Conversely, we can push another button, which controls another place in the brain, and cause intense punishment (fear, anger, pain, nausea, vomiting, unconsciousness, etc.). So we now have push-button control of the experience of specific emotion by animals in whose brains we have placed wires in the proper places." When an electrical jolt is transmitted along the wire, the animals "apparently can experience, among others, the following sensations: 'as if fed' when hungry; 'as if warm' when cold; 'as if cooled off' when hot; 'as if drinking' when thirsty; 'as if sexually approached' when deprived."

By putting the wires in lateral forward parts of the brain, said Lilly, lessons were imparted—the dolphin made those humanoid sounds the researchers were looking for. As we have already noted, although recognizable words were sometimes formed, the dolphins did not reach the stage of the later singing chimpanzees where phrases and sentences became evident. But the dolphins *may have understood much more* than the chimpanzees and gorillas, and Lilly suspected that they did. The dolphin's communication mechanisms were designed for under water, however, and those of humans for above water, a basic problem that has not been solved to this day.

Unfortunately for the interspecies cause, while the human side of the dialogue is meaningful (and a bit like a kindergarten teacher trying to perk up the class), the porpoise replies, in Lilly's report, consist of untranslatable buzzings, whistles, clicklings. What does seem conversational—especially when Lilly is past the electrode stage and letting human and dolphin have an interchange without artificial stimulation—is the back-and-forth nature of the exchanges. Human talks; porpoise clicks; human talks again; porpoise buzzes, whistles, gestures. It has the air of dialogue, all right.

The transformation of Dr. Lilly from the detached observer who manipulates the natural world with push-button technology to the concerned researcher who gathers data through personal involvement and interaction with his subject was a whirlwind odyssey that he dramatized to the hilt—and which deserved the dramatization, at that. It also led to a new approach to human-dolphin communication. The new approach aimed at simulating the mother-child relationship as a learning model by providing a live-in, sleep-in environment in which human and dolphin could be immersed in each other's worlds as fully as possible. In *The Mind of the Dolphin* Lilly writes that they hoped to expose the dolphin "to all of those actions, reactions, contexts, situations, and emotions which lead eventually to the baby becoming a child who speaks English (or other human language in the repertoire of the mother)." (As we will see, this approach has parallels in the ape-language experiments.) In addition to encouraging the production of humanoid noises and even words, it was thought that the live-in arrangement might also bring out other nonvocal modes of communication.

The human who took on the assignment was Margaret Howe, who joined the Lilly team in St. Thomas in February, 1964. Her previous experience with dolphins consisted of one visit to a dolphin circus in Florida. Her living companion was a young male dolphin named Peter. Although Ms. Howe started from scratch, she soon became a principal recorder of human-dolphin interchanges, and at times she was in full charge of the lab in St. Thomas. Lilly trusted her. Her willingness to undertake the particular experiment indicates her strongheartedness.

The facilities included a "wet room" flooded with twenty-two inches of sea water—shallow enough to allow the human to conduct her daily routine wading about from bed to desk to telephone, yet deep enough for the dolphin to swim freely and accompany her. Ms. Howe's sleeping quarters consisted of a foam mattress on a raised platform that extended into the flooded room.

The experiment, begun in June, 1965, lasted two and a half months, during which time Peter and Margaret shared each other's company twenty-four hours a day, eating sleeping, playing, learning games, working on vocalizing, relaxing, doing chores, taking showers, and so forth. On Saturdays Ms. Howe was allowed to escape.

During the fifth week of the experiment, Ms. Howe made the following report: "To actually live with a dolphin twenty-four hours a day is a very taxing situation. Much more so than I had anticipated. Unlike a dog, unlike a cat, unlike a human, a dolphin is more like a shadow than a roommate. If given the opportunity, *he will never leave your physical being.* To try and sweep a floor *with*

Peter, means that Peter is continually at your feet . . . touching you . . . pushing you . . . nibbling you . . . perhaps speaking (humanoid or Delphinese) to you. *He does not go away.* To cross a room to answer a phone means that Peter meets you when you come into his immediate range and he walks with you, pushing, nibbling, slapping, *the whole way.* and If you are on the phone for half an hour, Peter does not get distracted or bored, *he stays right with you . . .* again touching you, pushing, you, nibbling you, speaking, squirting."

The live-in, sleep-in arrangement had its risks. A porpoise can injure a human if both aren't careful. If complex language had already existed, it might have been possible to explain to Peter that he was not to be sexually aggressive with his companion. As it was, he had to be told by gestures, and there are some gestures that dolphins resist and ignore. Peter's sparky ways with Margaret Howe were not unique. The young women who swim with porpoises in oceanariums are familiar with the dolphin's sexual aggressiveness and often have to leave the pool to lower the dolphin's temperature. Margaret, who was supposed to determine how far communication could go if the cetacean and human became dependent on each other's thoughts, had no intention of surrendering to Peter, but she also wasn't going to cut and run because the dolphin acted like a dolphin. She wanted to hang in there as long as possible.

Next to Margaret in a shallow pool, Peter's emotional exchange began in the sexy interplay that has made the Howe experiment memorable to those who wonder what happens when humans and dolphins get together. At first, Peter conducted aquatic courtship, and the experimenter tried to figure a way out without destroying the premise of the experiment. Later, Peter softened his sexual advances into something more acceptable to the human, and Margaret Howe—caught up in an experiment that was gloom-making in its hardships—found herself discovering solace, friendship, and understanding in the dolphin. The diary she kept during this time is a document that only a world highly determined to be on the path toward whalespeak could have produced. That Peter adjusted his behavior, from the beginning, to suit the comparative vulnerability of his human friend is apparent. Before she went to live in the "wet room" she had seen Peter give a female dolphin,

Sissy, a bite in the eye. "He puts his head under her genital region and bumps upward hard," Margaret reported, "hearing many *whoos* as Peter move[d] in on Sissy in tight circles." Margaret Howe's dolphin diary indicates a changing relationship that progressed toward remarkable mutual trust. These excerpts indicate how it came about.

> I find that [Peter's] desires are hindering our relationship. I can play with him for just so long now, and then he gets an erection. . . . When Peter was upstairs in the Fiberglass tank he would occasionally become aroused, and I found that by taking his penis in my hand and letting him jam himself against me he would reach some sort of orgasm, mouth open, eyes closed, body shaking, then his penis would relax and withdraw. He would repeat this maybe two or three times and then his erection would stop and he seemed satisfied. . . . Now, however, I am completely in the water with him and because so much of my body is exposed, we cannot get into the same position as above. I am completely vulnerable to him and he pushes and shoves my legs and feet, and quite pathetically tries to satisfy himself.
>
> I have decided that Peter must go downstairs with Pam and Sissy [female dolphins] for at least a day. . . . This, I hope, will relieve his frustration.

> He listens, repeats, listens again. He has that lack of pronunciation, but improves daily on inflection and pitch. . . . In the middle of a cocktail party it could be considered background conversation. It has all the right "feel" of English . . . and soon it will be.

> Peter is courting me . . . or something very similar! . . . I compliment him vocally, soothingly, and rub him as he turns to be stroked.

> I started out afraid of Peter's mouth and afraid of Peter's sex. It had taken Peter about two months to teach me, and me about two months to learn, that I am free to involve myself completely with both. . . . Peter could bite me in two. *So he has taught me that I can trust him.*

> [*Feeling sorry for herself at the hardships of life in the wet room where she takes a shower in knee-deep sea water and wades back to bed, Margaret finds the dolphin is now as much confidant and sympathizer as wooer.*] I seldom if ever ignored him, and usually ended up right back in the water, not caring at all about sleep, or the wet bed or the shower routine . . . simply overwhelmed at what Peter and I were accomplishing together. . . . The feelings of depression and aloneness were not a constant thing by any means, but they did come and go, and my having to turn to Peter to overcome them was, I feel, an important part of the experiment.

The linguistic aspects of the experiment produced far more mixed results. While Peter did not learn to phrase in translatable (humanoid) English, he was at all times closer to human speech than any trainer had ever been to dolphin speech. Margaret, simplifying her name to "Magrit," could almost bring this from Peter in a form that all would understand. The "M" would seem to form and blend into humanoid sounds that mimicked the number of syllables, pitch, and inflection with a high degree of accuracy but not the actual enunciation of those syllables.

Lilly, looking at the taped record, concludes: "When he is wrong, Margaret can hush him and start over. . . . Peter improves in giving back the same number of sounds that are given to him. . . . Peter is able to make parts of his words understandable." He could say parts of *bobo,* parts of *clown,* parts of *good boy.* But water creature and land creature could meet emotionally far more easily than they could meet linguistically. "Margaret syncopates *baby block* and *basketball* and Peter learns to follow this," Lilly observes. "The lesson is controlled and formal, and the give-and-take of learning, teaching, speaking, and listening is established so that progress can be seen."

Lilly decided that the experiment had been "eminently worthwhile," and noted of Margaret Howe that "her intraspecies needs finally are being taken care of; she, like the woman with the chimpanzees in Africa, married her photographer."

Lilly sized up the essential barrier to translation and actual vocal communication in terms of the dolphin's basically acoustical orientation to the world, which is at present incompatible with the visual orientation of humans and vice versa. Writing in *The Dyadic Cyclone*, he notes:

> Their visual system is one-tenth the speed of ours; however, they make up for this in that their sonic and acoustic systems are ten times the speed of ours. This means that the dolphins can absorb through their ears the same amount of information—and at the same speed—that we do with our eyes. . . .
>
> Our visual orientation is built into our language so that we, in general, talk as if we were watching and seeing and analyzing what we were talking about as if *seen.* In contrast, the dolphins "see" with their sound-emitting apparatus and the echoes from the surrounding objects underwater, Remember that half the twenty-four-hour day, during the night, their eyes do not need to function. Remember that they must be able to "see" underwater in the murky depths during the day as well as during the night. They must be able to detect their enemies, the sharks; they must be able to detect the fish that they eat, and they must be able to detect one another in spite of a lack of light; therefore, they have an active processing mechanism for sound that is immensely complex.

DR. LILLY'S LARGER MESSAGE

Lilly's experiments—and they still go on—pointed us somewhere without taking us there. But Dr. Lilly had messages beyond the results of his experiments. In awe of no human, but considerably in awe of "the whale's giant computer," Lilly speculated from his own knowledge of whale brain, and from sea incidents reported throughout history, that whales are communicating with each other on a grand scale and that we merely deceive ourselves if we think they are limited creatures without a language.

"The whales know," Dr. Lilly would say in an apocalyptic manner quite different from his brisker style as a practicing scientist. "They know about World War I. They know about World War II. They know about submarine warfare. They recognize we have navies that could destroy them."

In contrast to the human species' warring ways, whales and dolphins—claims Lilly—routinely express benevolence toward members of their own species. They look out for one another, or, in Lilly's words, "they realize their total interdependence." Because whales and dolphins have no automatic breathing mechanism, they will drown if they become unconscious for any reason. Lilly reported many instances of fellow dolphins bringing an unconscious dolphin to the surface to restore breathing. As reported in *The Dyadic Cyclone*, "To wake one another up they will rake the dorsal fin across the anal/genital region causing a reflex contraction of the flukes, which lifts the endangered animal to the surface. Dolphins support one another at the surface and stimulate the unconscious one until the respiration starts again when he is awake."

Lilly's most important role may have been as the popularizer of the view that whales and porpoises also express this benevolence toward humans—often referred to as the "dolphin ethic" or the "whale ethic." Such benevolence is a deliberate choice made by reasoning creatures, Lilly states, intended to communicate goodwill and a thoughtful attitude toward fellow intelligent inhabitants of planet earth.

Although he could not produce a recording to show that the dolphins he tested had suddenly blossomed into full-blown human speech, Lilly believed there were endless incidents showing a complex, friendly nonvocal communication pattern in relations with himself and others in his team.

Elvar, the young porpoise who later seemed to "analyze human speech" as though he were going to do translations, did not begin in this brilliant fashion but had to learn lessons as he went along. Lilly recounts that when this very young dolphin once headed straight for him in attack position (despite frequent claims that dolphins will not attack a human), Lilly was able to escape because of the speed with which a mother dolphin intercepted Elvar and, Lilly explained in a seminar, slammed the young dolphin so hard that it literally bounced off the wall of the pool. "A great example of one-shot learning," said Dr. Lilly.

His point has always been that not only do porpoises learn but they learn a great deal, and the ethic they display is part of those other modes of communication that demonstrate language-learning as an ever-present possibility. Dr. Lilly argues that it is really a bit witless to believe that animals this complex don't exchange complicated ideas with each other and that they couldn't do the same with humans.

While there had once been a tendency to treat tales of dolphin *noblesse oblige* toward humans in distress as though they were mere superstition, those who were trying to relate the Lilly language theories to the deportment of dolphins in wild conditions had plentiful evidence that dolphins do indeed race to rescue humans in an emergency and are able to understand the situation precisely.

On September 10, 1972, the *New York Times* reported that Yvonne Vladislavich, a twenty-three-year-old South African, survived the sinking of a cabin cruiser that took the lives of three companions. The Indian Ocean off Mozambique where the cruiser sank is filled with sharks, and the sharks came for her—but dolphins intervened. Two of the dolphins, she said, fended off sharks attracted by blood from a cut on her foot. "They escorted her as she swam," reported the *Times*, "and helped her stay afloat when her strength was failing. . . . She said the two dolphins protected her until she reached a buoy and climbed up on it to await rescue."

To provide rescue help and standing escort of that kind is a large-scale act of communication, in the view of those supporting Lilly's theories. They also believe that a dolphin who can make such clever appraisals of a situation is surely not too dim to understand what connected language is all about.

ISO-HUMAN AND BEYOND

Dr. Lilly helped make it clear that the search for interspecies language is a search for powerful intelligence. Stressing the size and complexity of the brain as the physical basis for mental agility, Dr. Lilly finally indicated that the search for interspecies communication was really a search for intelligence at the seventh, eighth, or ninth level. He had worked out what those levels might be and defined them as follows:

Level 1—the single cell

Level 2—simpler invertebrates up through sharks

Level 3—birds, reptiles, some fishes

Level 4—low mammals up through the largest monkeys but not including the anthropoid apes

Level 5—orangutans, chimpanzees, gorillas—a special class of "super animals"

Level 6—proto-humanoids, those humans who are *barely* human, with an intelligence but low in intellect

Level 7—primitive human tribes with complex traditions and the beginnings of written language

Level 8—Iso-Humans (Dr. Lilly and most of us), civilized persons in society "with all its complexities . . . sciences, national relations, international relations, etc."

But beyond the eighth level, he saw Level 9, a category for "supra-Humans"—creatures with high intelligence but beyond us in intellect whom he suspected might be found in outer space. At times, he has seemed to suspect that whales may be at Level 9.

Among scientists, few have ever agreed with Dr. Lilly in imagining whales even to be within range of we Iso-Humans at Level 8. Most seem to believe that whales and dolphins belong at Level 5 with those "super animals," the higher apes.

Typically, as Lilly wrote, lectured, and expanded his ideas on why communication with whales and dolphins is entirely possible, he would state, "I invite you to entertain some new beliefs—these cetacea with huge brains are more intelligent than any man or woman." That seemed to be his final opinion—that they were at least at Level 8, going toward 9, and it was preposterous to imagine a creature of that intellect with not a shade of language.

ENIGMA, TEASE, AND PROPHET

By the late 1970s, Dr. Lilly had passed through many phases, many personal crises.

The later Dr. Lilly—the one his followers treat as the quintessential Dr. Lilly, a remarkable change from the implanter of electrodes—was very likely born in the mid-1960s and became apparent to a wide public at a dramatic moment in 1967. Five of the eight dolphins in the colony he was working with died at almost the same time, and Dr. Lilly diagnosed these deaths as suicide. Humans cannot commit suicide merely by holding their breath, but dolphins have a different anatomical structure and Dr. Lilly felt they had made a choice. (Most marine biologists disagree that this incident can be explained as suicide.) It was what Dr. Lilly did after these five deaths, however, that characterized him to a growing fleet of fans: he set the other three dolphins free. "I no longer wanted to run a concentration camp for my friends," he wrote. No other remark has been so widely quoted in relation to the long efforts to find some way of expressing a human ethic that would match the reputed benevolence of the porpoise ethic.

Dr. Lilly left porpoise experimentation for a time, working with himself and various others on altered states of consciousness. He popularized isolation tanks as a way of reaching some deeper self. He has run the gamut from rapture to despair and has always been willing to test his own deeper self to the limit. If I make Lilly sound all too close to lugubrious, put the thought aside—his wit comes like a whipcrack. When that arrowhead chin is lifted and his intellect is at work, Lilly probably comes as close as anyone in America to reflecting the style of Bernard Shaw—playing with the art of paradox, unafraid of large-scale theoretical doctrines, tweaking the presumptions of fellow intellectuals. I remember the quickness of his answers when bands of the Lilly faithful listened to him at a weekend study session in San Francisco. I rose to describe a friend's idea on how an artificial porpoise could be invented and the feces trails of real porpoises associated with this, so that tuna pursued by the tuna fleet would follow the decoy dolphins instead of scrambling underneath real dolphins and endangering them with the tunamen's nets. Wasn't that a great idea?

What made me think, said Lilly, that the tuna would be dumb enough to fall for it?

As a debater on the intellectual strengths of dolphins and whales, Dr. Lilly remains embattled and consistent. As an investigator into the possibilities of humans breaking through ordinary modes of existence, he sometimes passes into territory where no colleague—and no other interspeciesist—chooses to join him. Questioned as to the likelihood of interplanetary travel, he blithely remarked that he had himself traveled between planets as a point of light. It is never totally clear when Dr. Lilly makes such statements whether he wants absolute belief, is challenging the listener's willingness to follow him to any extreme, plans an explanation out of quantum physics as his follow-up, or is simply encouraging a general exuberance for crashing through the barriers of conventional thinking.

The theories Dr. Lilly is willing to advance to stimulate a group of dolphinatics into visionary discussion drive the literal-minded to write him off as one of the higher con artists—the kind of public figure who, brilliant enough, will tell almost any tale to make sure that he, among all others, commands the most attention.

Dr. Lilly teases at times, but he always comes back to the data. His experiments with porpoises are purposeful and well recorded; he bluffs no results. His actual results have stopped short of proof of the interspecies theorem, but he knows this and continues to treat his own ideas as theories, not proven fact.

Although he has never demonstrated that dolphins talk, his connection to the debate goes on, for Dr. Lilly created an atmosphere in which millions are ready to believe in, and mightily welcome, the talking dolphin and the talking whale.

If the quest for interspecies communication seems so different from all the other quests of modern science, that's partly because Dr. Lilly has chosen to have it that way. He has proved to be one of those rare persons who believe that views of the universe are theirs to revise. In eighteenth-century France, in revolutionary America, in nineteenth-century England the world was full of such men, and we have lately, to our sorrow, lacked them. Dr. Lilly is exciting. It's one of the reasons the world bothers to listen. Somehow he has managed to combine the idea of how to be a scientist with how to be an orator, raising great questions like some new kind of Daniel Webster.

"Discovery, in my experience, requires disillusionment first as well as later," Dr. Lilly told an audience in 1962, slipping into the style that suits him best. "One must be shaken in one's basic beliefs before the discovery can penetrate one's mind sufficiently to be detected. A certain willingness to face censure, to be a maverick, to question one's own beliefs, to revise them, are obviously necessary. But what is not obvious is how to prepare one's own mind to receive the transmissions from the far side of the protective transparent wall separating each of us from the dark gulf of the unknown. Maybe we must realize that we are still babies in the universe, taking steps never before taken. Sometimes we reach out from our aloneness for someone else who may or may not exist. But at least we reach out, and it is gratifying to see our dolphins reach also, however primitively. They reach toward those of us who are willing to reach toward them."

John C. Lilly *described* experiments—he can be quite factual and conventional while doing so—but he *promised* victories. This was largely tactic. It was what kept the interspeciesists going in spite of predictions of disaster and defeat. If he is considered not as the primary architect of the proofs of interspecies communication but as the morale-builder whose visionary ideas and intellect have pulled restless minds into the effort to break a puzzle, then Lilly's performance over the past three decades—with all its prophecies, melancholies, great pronouncements, sudden retreats, and new pronouncements—becomes a perfectly intelligible and logical performance.

As recently as 1961, when Dr. Lilly began his prophecies, the interspecies movement was still a tin lizzie that needed some heavy cranking to get it going. Dr. Lilly cranked it up and it's still rumbling.

By 1980, Dr. Lilly was beginning to describe to interviewers the circumstances he thought might lead to fuller communication with dolphins. He spoke of wanting a seaside house where the creatures could come and go from the sea, independently. It would have a dry area for humans only, a shallow wet area where humans and dolphins could congregate together, and a deep-channel area providing an ocean route for the dolphin.

If whalespeak becomes reality, Dr. Lilly, working at a marineland near his home in Redwood City, California, will be among the first to know. And he still seems to expect that a whale may bring him this message personally.

4

Symbol Trips: The Musical Elephant In The Behaviorist's Lab

IN THE 1950s THE GERMAN INVESTIGATOR BERNARD Rensch, director of the University of Münster's Zoological Institute, set out to compare small brains to larger brains, and big brains to even bigger ones. As he worked his way up the scale—comparing the brains of mouse to rat, of dwarf Indian squirrel to giant Indian squirrel, of small chicken to large chicken—his destination became inevitable. The largest of land-brains belongs to the elephant, whose brain weighs in at thirteen pounds.

It is just possible that Rensch and the elephant shared something. Rensch was clearly a man of systems and, in its own way, the elephant can seem a systematizer, too. Out in the jungles of Southeast Asia, the elephant is easily taught to nudge great teakwood logs into organized stacks. The elephant not only has the strength to do this but also appears to understand symmetry.

Remember how you have gone to the circus and found what the promoters called their "five herds of elephants," (consisting of a total of ten animals altogether) forming perfect patterns with their ring-around-the-rosie circles, their mutual headstands, their triangles and apexes? In the jungle, their performance with the logs is at least as ingenious as it is at the circus. We could say that it is a long-existing example of interspecies communication going back many centuries, and few in India or Laos would quarrel with that. They know that elephants understand a wide range of human language. They also know that something else is going on among elephants: that thirteen-pound-brain is forever on the move—checking, calculating, planning, and (or so it is said) deceiving. A story in a book called *Elephant Bill* would, if true, classify the elephant as a creative thinker. The author, J. H. Williams, who had worked with elephants in Burma for a quarter of a century, claimed that these great lumbering creatures had plans for making off with bananas at night, plans that were foiled by the clanging bells they were forced to wear around their necks. And so, said Williams, the elephants would stuff mud into the bells to muffle the sound so they could go about their banana burglary without alerting their keepers.

Professor Rensch was himself intrigued with *Elephant Bill* but felt the stories credited elephants with "far too much insight into the future to be believable." If the claims in *Elephant Bill* seemed incredible to some, there was another remarkable elephant trait that could be easily studied and verified in the jungle. At the long slide, this creature who goes about its tasks with lumbering precision places each log neatly on the spillway so that it will speed smoothly into the water. Setting the log with its trunk, pushing it off with its forefeet, the elephant exhibits another kind of behavior that we can all recognize: it watches the log all the way down, with a perceiving eye, to see if it makes the right entry. Although elephants are not found in bowling alleys, they were clearly behaving very much like human bowlers. Trained to do massive chores, elephants also kept checking to make sure they had made good shots.

This creature with the creative brain represented a culmination for Bernard Rensch. Determining ability in language is the classical direction in which to head with an investigation into animal intelligence. It is entirely possible, however, to head there without arriving there. Although gorilla and dolphin investigators would achieve a great deal that Rensch did not, in terms of what he was trying for, Rensch arrived.

Taking the elephant on a symbol trip seems farfetched only because of the animal's size. As Rensch explained in his article "The Intelligence of Elephants" (*Scientific American*, February, 1957), he did not allow himself to be stymied simply because the elephant is "not particularly convenient as a 'laboratory animal.' " The zoo in Münster had a five-year-old female elephant who was about to make elephant history based solely on her proximity to the institute where Rensch had come in need of a large-brained animal. The sperm whale, a fifty-footer, edges out the elephant with a seventeen-pound brain in an immensely greater body, but sperm whales were not available. So Rensch settled for the elephant, intending to compare it with horses, asses, and zebras.

THE ELEPHANT'S "VISUAL VOCABULARY"

Rensch's language experiments are difficult to appreciate for their actual meaning unless you put yourself in a proper frame of mind to understand what is going on. For purposes of the following discussion, I suggest you think of the elephant as a poker player in a game where the cards are face up. The elephant wins the pot—that is, gets the reward—only if it picks the card that has been designated as the "right" card.

The elephants were tested with what amounts to a large-scale deck of symbolic cards, but in this case the cards were mounted as cardboard lids on small wooden boxes, used two at a time. In the jungle, it's possible that the elephant derives satisfaction from knowing that it has made a good shot, but in Professor Rensch's version of an elephant laboratory the conventional stimulus of "food if you choose right" was used to induce the desired behavior. The elephant could win a piece of bread by opening the lid with the prize behind it. In the beginning, one box was marked with a circle, the other with a cross. Because the elephant had a natural tendency to choose the box with the circle on top, the experimenters decided to designate a choice of the circle-box as "wrong" and a choice of the cross-box as "right."

"Nein!" shouted the Germans when the elephant chose the circle-box. In the first few days, the elephant learned to choose a

cross when she saw one—and with no further need for a yelling of "*Nein!*"—after 330 trials. To determine the subtlety of the elephant's comprehension, Rensch turned the cross to look like an X and found that the elephant could still make the correct choice. Then the experimenters began series of symbolic tests by introducing new designs of many sorts to the elephant. What were box lids to the elephant were "stimulus pairs" to Rensch and associates. To us, it's just the same as learning to know a jack from an ace. We say the ace is higher ("right") simply because we were taught to; it's the rule of the game—no other reason in the world.

In each stimulus pair, there was a "wrong" and a "right," and the elephant learned to keep all these in mind simultaneously through extensive tests to see how much symbolism could be held in the mind at one time.

"In a final multiple-choice test," reported Rensch, "each of the twenty pairs was presented thirty times according to a previously established sequence. The elephant mastered all twenty discrimination pairs superbly. In most of them she made only one or two wrong choices or no error at all. The test covered 600 trials lasting several hours, yet the elephant not only showed no symptoms of fatigue but actually improved in performance toward the end."

Perhaps the most memorable discovery in this phase of the Rensch testing—he concluded that the elephant could go much farther and "master a bigger vocabulary" but felt the point had been proved—lies in the elephant's reaction to being forced into an unsatisfactory choice. Departing from the idea of "right" and "wrong," the experimenters presented the elephant with "wrong" and "neutral" (a blank card with no sign on it).

"When she made a choice," said Rensch, "she would often take the neutral one. But sometimes she would become excited and tear or bite the cardboard lid or trample on it. Her behavior suggested the kind of experimental neurosis that has been produced by conflict situations in many other subjects from fish to man."

Rensch did not hesitate to call what the elephant had learned "a visual vocabulary," and the same idea obtains in a number of animal intelligence tests carried out with other creatures in other places. Such symbols are, in a sense, "words." The Rensch system was an extension on the word "*nein.*" The elephant knew the concept of "You're wrong—don't do that," and most animals can take that kind of interspecies message from a gesticulating human, heavy-handed with signs and shouts. But to carry out the idea of "You're wrong—don't do that" so that careful discriminations are made between long lines of symbols is almost like being able to perform the act of "naming."

What significance does a visual vocabulary have for teaching elephants, or any other animal, to talk? Think about the poker game; it becomes clearer. If poker players from half a dozen countries all spoke different languages but knew the basic rules of Stud, Draw, and Down the River, they could win or lose almost as handily as if they were all conversant in the same language. Why? Because aces talk in any language. The tongue they speak is international. Is it interspecies?

A suspicion arises that it could be. The cards that humans use probably seem too sporty to be scientific, so animals are generally not offered a choice of aces, kings, queens, and the like. But the principle remains the same. A child can learn to tell a 4 from a 3. An elephant, as Rensch showed, can do this, too. And it can go farther—considerably.

The elephant went from the idea of a difference between circles and crosses to carefully distinguishing the difference between wider stripes and narrower stripes, between three dots and four dots—and it was easier to see that these choices constituted a symbolic vocabulary. When they began to shuffle the three dot patterns and the four-dot patterns, there was a preliminary difficulty for the elephant. But she broke through it. No matter how the dots were arranged, she could choose between the three dots ("right") and the four dots ("wrong"). It was as though the elephant had learned to say, "Okay, I know three dots when I see them and that's what you're driving at. What's next?"

What's next was determining whether the elephant could hold onto what she had learned for a length of time.

"We proceeded to a test of the elephant's memory with very interesting results," Rensch proclaimed. "She was presented with thirteen pairs of cards which she had learned earlier but had not seen for about a year. In a total of 520 trials she scored between 73 and 100 per cent on all the pairs except one, which was a difficult discrimination problem (i.e., the double circle vs. the double half-circle, on which she scored 62 per cent). In other words, the elephant had retained the meaning of different visual patterns for the period of about one year."

Next came a music lesson for the elephant. Photographs of how these lessons were carried out suggest that the conditions were less than ideal. They show an elephant behind bars, reaching outside the cage with her trunk for a switchbox. If she reached when the "right sound" had echoed in her cage, the gadget automatically put a chunk of bread within her reach. If a child were forced into a music lesson with heavy bars encumbering every movement, the caterwauling might be great. Again, with the prospect of food if she reached for the switchbox on the right note, the elephant cooperated.

"The elephant learned to discriminate six pairs of sounds," reported Professor Rensch, "one of which differed by only a single full note. In tests on all of them in irregular rotation, she was able to distinguish all twelve tones and to know their positive or negative meaning."

This experiment was carried out with an investigator concealed behind a screen watching in mirrors to see how the elephant performed from the sounds of the tones alone. Rensch thought the experimental results showed "an excellent memory of absolute pitch." Again, they waited a year to check the elephant's memory and found that she could still pick the right pitch "in nine out of twelve tries."

We will probably always be troubled by the thought: *But the elephant is sparkier than this at the circus any day.* True enough. The notion that choosing between several sets of tones is the elephant's all-time high in musical achievement can seem like nonsense to anyone who has watched elephants amid the strut and troubadouring of the Big Top, where they react to music in many and exotic ways. Still, it was their systematic nature that made the Rensch experiments reassuring to those who want their science from a scientific source.

Music, produced from a system enchanting in its perfect symmetry, would appear to be a field for further exploration with the orderly elephant, a creature capable of precise dicriminations. Even though it begins in symmetry, music goes on toward wild flights of absolute creation. How wild a flight could the elephant take if it were given a chance to choose not just notes but symphonies of sound from anywhere in music or anywhere in nature?

We don't know yet. It's possible we don't know because our experiments are too laborious in their methodology—have not yet taken flight themselves. We have introduced the animal to music in a limited way and to card playing in a limited way. The elephant learns forty visual designs in its card playing, while children learn thirteen, but children are given more intricate fun. Progressively, we learn to shuffle, deal, draw to an inside straight, bid four hearts, go to a tournament. Animals begin the experience, but it isn't progressive. Because the animals are so slow? Or because we don't know how to make the experiment reflect a complicated experience?

The behavioral researchers gave the elephant a few stray experiences in music and card playing but the big stuff—the stuff that makes music everlasting with us and card playing a favorite sport—is nowhere to be found. The behaviorist is failing to convey glory and suspense; the experiment fails to tingle the spine.

Often, experimenters seemed to find no way to give their symbol trips some larger dimension for the animals. Though Rensch unquestionably felt that his experiment went farther than any other, the circumstances raise a question as to how far it could go. Is an elephant behind bars truly an elephant? There is a bit of scientific opinion to the effect that it is not. In 1923, Wolfgang Köhler, a theorist in the field, had suggested that "a chimpanzee kept in solitude is not a real chimpanzee at all."

We cannot hope to discover animal depth and complexity if we treat them as though they are simple. A fallacy of sorts can be sensed from the Rensch experiment. The questions for the elephant change and grow harder, but the reward is always prosaic. For an expert performance on a long multiple-choice quiz, the elephant is still just fishing up the same old bread chunks. If hunks of food disclosed in a box like crackerjack prizes are an elephant's sole idea of the satisfactions to be found in life, then we cannot regard it as having much intelligence, regardless of how dextrous its performance on tests. Brain implants such as Lilly used are hardly the answer to this dilemma, either. Depriving a newly caught wild animal of food to break its spirit and make it come begging to the human keeper for something to eat is a technique that has been used by trainers. It produces communication of a sort, but it also seems guaranteed to put the relationship on a ruler-and-beggar footing.

The interspecies communicators of the last two decades have made a departure, for they tended to bring with them not only a more ambitious goal but a determination that animals have to be treated bountifully if they are to display the most personable traits.

Remember Strongheart, the movie dog? Boone claimed that Strongheart's idea of a great satisfaction was to head for the high country, bounding along (with Boone trying to keep up) as though some great glory awaited them if they could go far enough and high enough. Arriving at a high vantage point in the mountains, the dog would stare off into the sunset, stare into illimitable space for a long, profound period before it was ready to go home again. Boone treated this as a soulful act, and yet reasoning of that kind deeply disturbs the investigator who wants to explore animal intellect in quantitatively measurable terms and believes that any hint of mystery must be avoided at all costs. This would be an easier premise to establish in regard to the intelligent animal if humans themselves were not so mysterious and if each of us were not mysterious to ourselves. If we relate Strongheart's high-country yearnings to our own experience, we know there are unconscious stirrings that can govern our own sense of where to go and when to go there. Who has not run to the high country, now and then, when life below seemed all too stifling? As humans, we can take satisfaction in such a journey, and we find satisfaction in many things besides food; anyone who believes that our inner yearnings can be taken care of with the offer of a candy bar or a Coke is apt to inspire contempt.

Why be so certain that animals are otherwise?

If we can argue, from the Rensch experiments, that elephants are still at an elementary stage of utilizing symbols as a "visual language," we should also concede that the "rewarders" are still at an elementary stage with their rewards.

We can't be sure how comfortable elephants feel with symbols or whether symbols spark their interest. Professor Rensch felt that the elephant he worked with became quite absorbed in the attempt to beat the guessing game of the cards. Our usual presumption, as humans, has been that elephants grow precise and thoughtful only under human prodding. Yet Plutarch reported, from the first century A.D., that an elephant who thought itself alone was seen practicing its dancing steps just like any young human caught up in the local theatricals.

THE TALKING HORSE REVISITED

How much can an animal communicate when it has a deep desire in its heart or even a deep aversion? How well does it do this and in how complicated a fashion?

The best answer to this lies outside the behaviorist's laboratory. What Rensch discovered about elephants, with symbols and switchboxes, was comparatively small, valuable only because it offered acceptable scientific backup for traits that can be glimpsed far more readily in the animals' everyday lives.

When he went to India, Rensch discovered elephants understood the difference in commands as similar as "Lie down on your belly" and "Lie down on your side." This was not so different from the elephant's ability to distinguish between narrow lines and wide lines, between three dots and four dots. "Lift up your trunk," "Duck under the water," "Push with your foot," "Push with your head"—elephants had, for generations, been perfectly cognizant of the meanings of all these phrases. Once he had established the scientific basis for the kind of elephant discriminations that all the elephant handlers in India had been aware of for generations, Rensch set up comparison tests with an ass, a zebra, and a horse. That was simply to follow through on his original *big brain/smaller brain* comparisons. Ass, zebra, and horse were compared with the elephant in their ability to interpret symbol-pairs.

The zebra, it developed, could master only half of the twenty stimulus pairs that the elephant had understood in their entirety. The ass mastered thirteen. But the horse was able to learn all twenty pairs—"Surprisingly enough," commented Professor Rensch. He hadn't expected the horse to perform that well.

This puts a stronger, more scientific light on what the English horse breeder Henry Blake was contending, in *Talking with Horses,* when we met him in Chapter Two. Blake had a number of strong ideas about horse intelligence, one of which came quite close to what Rensch was establishing with his symbol-pairs. Blake was trying to show—and inviting the reader to find the proof in his own pasture among his own horses—that horses do in fact make delicate lingual distinctions, not just general snorts and tail wavings.

Blake was interested in how horses announce their feelings, but he also took the signs of extensive vocabulary as reflecting an animal psyche capable of deeper and more inventive thinking than it is customary for humans to admit or behaviorists to probe for.

Among the horse expressions in his "dictionary" are such carefully delineated expressions (if you can accept their reality) as: "Mummy loves you," "You will be quite safe with me," and "We are good girls here."

How deep a symbol trip are horses really on, then? How far can it go? How much reach does their "language" have and what deeper thoughts are there in the brains of horses other than the familiar desire for food, running room, and an occasional neck rub?

Blake offered a sexist finding about stallions and mares. Nearly all the stallions' utterances, he declared, have to do with three subjects: sex, danger, and food. The mare, he concluded, has wider concerns, which would seem to derive from motherhood. He finds mares much concerned with ideas related to the care of foals and conveying this concern in signs that humans or horses can easily read.

Blake's best story in *Talking with Horses* concerns a horse named Fearless who changes from murderer to demonstrative friend in one lifetime. If we accept that a "great moment in inter-species communication" is a least as likely to occur between an animal and its owner as between a scientist and subject, then I would nominate the homecoming scene Blake describes between himself and Fearless. Although it's a bit more earthshaking than the usual reunion between a horse and its owner, many a ranch kid may find that he has experienced something similar and knows what Blake is talking about. To Henry Blake, it was an example of a horse talking with exuberant expressiveness.

When the horse breeder experienced this great moment, he had just returned to the pastures of Somerset, England, after an overseas stint with the Army.

"Come on, my darlings," he cried to the horses stretched out across a ten-acre field. All pricked up their ears and came flying to meet him, Fearless in the lead. Fearless had killed her previous owner and had begun her stay with Blake by trying to chomp chunks out of his legs and return his friendly overtures with kicks. Despite this, the horse had become his great friend—"She saved me twice from injury and on one occasion saved my father's life." Horses, according to Blake's dictionary, certainly know how to say "Welcome," and here's how Fearless did it when Blake came back after his long absence:

> She was galloping as fast as she could with her head stuck out and her ears flat back and her mouth open. Even after eight years she was still liable to have a piece out of you or to clout you with her front feet, and since I had gone too far down the field to make a run for it, I stood still. When she got to within ten yards of me she stuck her four feet into the ground and skidded to a stop. Then she took two steps forward and licked me all over from head to foot, and when she had done this for about three minutes the tears were running down my cheeks, so she thought that was enough of a good thing. Just to show me the status was still quo, she caught hold of me with her teeth and lifted me from the ground and shook me slowly backward and forward four or five times, then she put me down and rubbed me with her nose. I have never been so touched in all my life, the display of affection was somewhat unusual in form, but it was fantastic.

Rensch showed that horses, sometimes portrayed as remarkably stupid, actually have a fine grasp of symbolism even for such laboratory stunts as card-reading. Blake provides anecdotes by the score. But how deep is the brain of a horse, really?

Of all Blake's tales on the ability of a horse both to understand human instructions and to go against them when it chooses, the most telling is this: He suggests that the real brainwave among horses is the one who knows how to ignore human instructions and learns to throw a race—deliberately. The *truly* smart horse, according to Blake, is too smart to fall for its owner's conniving efforts toward making it (a hollow victory for a horse) a racing champion. "On the race course they [the smart ones] quickly learn that there is no profit in racing, especially in a tight finish, when they will get a hiding, and so they pack in racing and become dogs."

The Rensch experiments need to be followed up on because they imply great intelligence and considerable dexterity with symbol-language in both elephants and horses. But the comparatively more elaborate reports from elephant trainers and horse breeders—the feats that are reported from the jungle and the cow pasture that are far greater than those from the behaviorist's lab—suggest that it will be the horse-naturalist and the elephant-naturalist, watching these animals with their own kind, who will bring us the most profound information in the future.

That's true, at least, unless some future Professor Rensch can show an elephant how to shuffle the deck.

5

Symbol Trips: The Dog, the Typewriter, and the Critic

ELISABETH MANN BORGESE, PET LOVER AND THE youngest daughter of novelist Thomas Mann, was an assiduous collector of every scrap of research and information on animal intelligence and the possibilities for interspecies communication. She took a firm interest in the reports coming from Germany about Rensch and his elephant experiment.

Others had heard about the elephants, too, but Mrs. Borgese had a streak of ingrained curiosity that caused her to take action where others sat and wondered. Her reaction to the reports by Professor Rensch was to go to Kerala State in South India and check out elephant intelligence personally. She found it just as remarkable as Rensch had indicated and came back with a load of stories on the mental agility of elephants.

Her most memorable contribution in animal communication was her notion that it might be possible to lead a dog to use a human vocabulary. Based on her studies, she was willing to speculate that the animal brain can come closer than we might think to mastering language. If the physical means to send the human a message were within reach, she reasoned, a dog might carry this far beyond the possiblities in a well-spoken bark.

She believed the animal would need a way to handle letters and make words when nothing in its physical makeup was adapted to that purpose. She boiled the problem down to anagrams. How could the game of anagrams be made simple enough to enable an animal—say, one of her own pet dogs, if it had the inherent intelligence—to begin to get its ideas across?

She settled on an Olivetti typewriter specially designed to dog specifications in order to test out her hypothesis. While Trixie, Jinxy, Pluto, and Arli all participated in her early experiments, it was Arli, an English setter, who leapt to the head of the class and became the canine typewriter champion of the 1960s.

Mrs. Borgese's work with Arli, which we will go into, was a novel follow-up to the clinical work with animals dating back to Pavlov and a number of others. Quite as important as anything she may have established about animal intelligence was Mrs. Borgese's willingness to believe that an experiment could be developed along scientific lines with frisky animals whose intelligence had been freely developed in an ordinary family situation.

Following are some of the developments that gave Mrs. Borgese hope for success with her experiment:

> Pavlov had made a remarkable claim, quite similar to what Rensch proved about elephants, concerning the discriminatory power of dogs. He asserted that dogs can identify the single notes in a chord and can differentiate quarter tones with precision.
>
> Dr. Otto Koehler, an editor with Konrad Lorenz of the *Journal of Animal Psychology* and a University of Freiburg zoologist, was reporting successes in teaching ravens, Amazon parrots, magpies, and squirrels to accomplish "nonverbal" counting. A jackdaw had been taught that if it saw two dots in a circle, it meant two mealworms were available for eating—that many and no more. If it saw four dots, it would search to find the four mealworms, stopping when it had fulfilled the count.
>
> The Russian Nadie Koht, working with a chimpanzee to see if it could make the connection between sight and touch,

put blocks of assorted shapes into a bag and then displayed in her hand the shape the chimpanzee was to pull from the bag without looking. The chimp's success in performing this task indicated what psychologists called a "hetermodal transposition"—discriminating between senses. It was relating sight and touch. Humans make such transpositions without even thinking about it, but there had been some doubt whether animals did.

A gray parrot named Jakob, as reported by Koehler, worked up to a different and even more bizarre version of the "hetermodal transposition." Jakob looked for two food bits on a count of two, three on a count of three, and so on—but did it by sound. A recorder provided four successive sounds—Jakob would count until he had found the four food bits and then stop. The sounds were then replaced by light flashes. Jakob still made the connection. When sounds and light flashes were made in combination, he didn't fly into a nervous breakdown (as the researchers had every right to expect) but just kept on counting efficiently.

Koehler summed up the meaning of these and other experiments, noting that the important element here was the animal's power to grasp the *idea* of naming and of counting even if it didn't have actual names and numbers to go by. Although Koehler denied that animals can use language in the human sense, he made a statement that still seems like a revelation to an experimenter bent on proving the interspecies theorem that humans and animals can exchange their thoughts. "Thorough examination of all the . . . counting experiments dating back half a century teaches the same lesson," wrote Koehler in an essay. "So far as we know, adult human beings are in no way superior to animals in nonverbal thinking, where matters that actually concern animals are involved. . . . Animals cannot name but they possess the power of nonverbal thinking."

Mrs. Borgese's typewriter experiment had a charming simplicity. If the animal had thinking power, naming power did not seem so far off. What are numbers but names—names for a concept that even the jackdaw could grasp. Mrs. Borgese looked at the Nadie Koht experiment with the chimpanzee and saw that the chimpanzee must know what a triangle was; it just didn't have the name for it. What it did have, she asserted, was a sense of "the idea of triangliness."

This question of naming power can rise to plague the interspecies communicator. It sounds ephemeral somehow, but, after many misfortunes in imagining that nonverbal thinking is a close relative of verbal thinking, the problem began to focus. It's possible that the name for an idea, surely not *more* important than the idea, is nevertheless vital. Every achievement in human life starts with a symbol trip. The spaceship begins as an idea; the idea can be transmitted to others only by naming and describing it; in the course of explaining, ideas on how to build such a spaceship are piled aboard. Thinking may be the motor, but verbalizing becomes a drive shaft.

Koehler had prefaced his bold statement on animals' power of nonverbal thinking with a careful modification. They have that power, he said, *on matters that actually concern animals.* Even with this proviso, Mrs. Borgese could find many reasons to believe that animals could move into significant new territory if she simply provided the means for transposing into words the ideas a decently intelligent dog might already understand.

A typewriter seemed exactly the way to go if you could teach a dog to use it.

Mrs. Borgese's intention was to lead the dogs by careful stages into an understanding of how the machine operated. She was able to start on this while the machine was still being built.

First, she offered food rewards to cause the dogs to nose at saucers, and then she switched them to nosing at saucer-shaped typewriter keys. By leading them to understand symbols (nosing the key with the right symbol on it) the dog might, she reasoned, become a typist. How much he would understand of the process by which a key, properly pressed, plants the same symbol on a typing sheet was in doubt, but she was determined to explore it.

The symbols on the Olivetti were a combination of numbers, letters, and geometrical signs. Two and a half inches in diameter, the keys were designed for snout, not for paw. It brought the dogs' eyes close to their work.

Imagining the process by which the dogs might be led from food saucers to typewriter "keys," Mrs. Borgese started a patient regimen that was quite as inventive—and used many of the same principles—as the techniques of the behaviorists she had studied.

No date has been recorded for the advent of the human ability to write, but if it should be agreed, in the future, that dogs can write on a typewriter to some significant purpose, we know from Mrs. Borgese that the dogs started on marked saucers on October 25, 1962. The most successful of them, Arli, "was introduced to his typewriter on July 10, 1963."

At no time did Mrs. Borgese seem to imagine that typing was possible for all dogs up and down the block. Jinxy and Pluto were started on much the same basis as Arli but lost out on the basis of not as much talent, and even clever Trixie fell by the wayside. These other dogs, Mrs. Borgese reported, "accepted the injustice of life; Arli went to school, and they did not." Their teacher realized that what she had invented was a "thinking dog's course of studies" and that it was "tailored to suit a genius."

A genius she believed she had—Arli—not a genius in the human sense but a genius among dogs. A setter with a half-black, half-white face, he had been christened Arlecchino (in Italian, "harlequin") and humans—who not only have names but *names for names*—quickly made him into Arli.

Mrs. Borgese is probably the first teacher of typewriting to hold raw hamburger under the letter A in order to get her student started. It should be remembered in what follows that Arli was not being invited to nose his way into a creative run of sentences from the first moment. Like any other typing student, he was instructed by dictation. Since he could not read a typing book, Mrs. Borgese dictated aloud. In the preliminary round of symbol training, she had taught him to nose the symbols for CAR and CAT and ARLI and also for 1, 2, 3, 4, and 5. The complicated training system, using saucers sometimes smeared with ham fat but always marked with identifiable symbols, is described in detail in Mrs. Borgese's short, hopeful book, *The Language Barrier: Beasts and Men* (1965). Like Robinson Crusoe carefully marking his homemade calendar, Arli's trainer was careful to date every advance.

Ten days after Arli first put his nostrils to the typewriter, he was nosing out those words, including his own name, that he had learned earlier. Within six months, his vocabulary had been enlarged to include PLUTO, BONE, GO, BAD, and a number of other words, including ROME. (One can speculate what Arli might have understood from the word *Rome;* most of his vocabulary, though, was within the range of names that dogs often learn to associate with familiar subjects.)

Several years of busy typing followed; Mrs. Borgese kept the typescripts. By March, 1965, the world's first English-setter typist recognized (by Mrs. Borgese's count) eighteen different letters and had a typing vocabulary estimated, at its top, at sixty words. Arli never saw the connection between hitting the keys in sequence and the appearance of words on the paper rolled into the typewriter platen. This part of Mrs. Borgese's report is a loose end, and a serious one. It seems to point to a dog who can write but not read.

Two of the incidents from Arli's training period are more significant than others. Usually Arli was no more than a transcriber—sometimes a mixer-upper—of words that Mrs. Borgese put to him. But the question, "Arli, where do you want to go?" excited him so greatly that he had an answer. Arli's desire was to go for a car ride and, as Mrs. Borgese recounts it, his breakthrough into communication of his own came with this question. He got so excited he became a typewriter stammerer, writing ACCACCAAAARR and GGOGO CAARR. His usual typing was not immaculate, either, but he generally didn't stutter to this extent, and sometimes, words were spelled just right (BONE GOOD . . . EGG BAD . . . SMALL TALL BALL MALL). A second unusual incident, which Mrs. Borgese was to ponder about as the closest she and Arli came to that "verge of the breakthrough to real communication," led to a change in her method of usually dictating what he would write.

Arli's unusual communication came about during a stress period. Landing at Santa Barbara, California, after an air trip, he had the canine equivalent of jet lag, and his nerves and stomach were both reacting. He couldn't seem to settle down, staying troubled over a considerable period. To try and increase his enthusiasm for going to the typewriter, Mrs. Borgese dictated, "G-O-O-D-D-O-G-G-E-T-B-O-N-E," and Arli, putting spaces in—for he had learned to use the space bar—responded "A BAD A BAD DOOG."

Mrs. Borgese was excited—and yet it seemed to be a dead end. When the incident did not repeat itself or flower into new and ever more revealing exchanges, she wondered if it was only by chance that Arli had hit on familiar words because he hadn't been inclined to follow her orders on typing.

And so she decided, "Why not let him type *without* dictation? Why not let him type whatever passed through his mind (or nose)? This experiment in spontaneous writing was initiated on March 21, 1965."

She does not say how Arli was induced to type without her dictating to guide him, but she produces, at the end of her discussion, a final run of Arli scribble that is described as purely Arli's own. It begins: ". . . AT ARLI GO A A A CAT A CA AND E E DSEE A BED DG DOOG GO BAD DOG BE AN GOOD DG DOG AD D A AND A GEU O OGET BD BALL AD AND GO BED ARRLI CA"T RIA RIVAL RIA C RIB RIVAL RIVAL RIVA"L RI VATL RICVAL ARLI EAT EGG AND MILA MILC MILC . . ."

Not a triumph of clear communication, but Arli comes back, amid the garble, to words that he knows and even ideas that may be important to him ("good dg dog").

During the early training period, Mrs. Borgese had said, "Some days the trainer is under the impression that the animal really has learned nothing at all and that any positive results are pure chance. On other days the animal seems to understand everything. One goes on week after week, month after month, year after year, with the tenacity and hopefulness of the alchemist of olden days, who repeated his experiment with the same imperturable mixture of calm, resignation, and faith, through the decades, through a lifetime."

Now, a bit like the alchemists—since she hadn't found purest gold in what Arli had to say when on his own—she began to search through the new spontaneous utterances to see if there might be more there than met the eye. She took the Arli typescripts, which ran in continuous lines, and divided the phrases up a bit in the style of e.e. cummings, the lowercase poet, and achieved a kind of doggerel. One brief quote, since it resembles all the rest, suffices:

CAD A BAF
BDD AF DFF
ART AD
ABD AD ARRLI
BED A CCAT

Does Arli really mean *cad* when he says *cad*? This seems doubtful, but even if we count this, there are five recognizable words here in a mishmash of fourteen, most of no meaning. Mrs. Borgese was incautious enough to repeat rather proudly that "a well-known critic of modern poetry" had decided that these meanderings from Arli's snout were poems of definite charm. "I think he has a definite affinity with the 'concretist' groups in Brazil, Scotland, and Germany," she quoted the critic as saying. "Has he been in touch with them?"

The urge to present Arli as possible poet should have been resisted. If it had been, the rather carefully wrought experiments from the setter's early period might have some firmer place in present-day recapitulations of "the talking animals."

Thomas A. Sebeok, a professor of linguistics and chairman of the Research Center for Language at Indiana University, dismissed the Borgese experiments out of hand. A serious, highly credentialed critic of the interspecies experiments, Sebeok has been a leading advocate of the view that only humans can have language and often casts himself in the role of combatting modern superstition.

He dealt with Mrs. Borgese in cavalier fashion. In the oddly titled article, "Talking Dogs" (odd because its intention was to show that dogs do not talk), Sebeok declared that all reports of talking dogs "fall into four distinct categories." As he defined them, these four categories were: mythological talking dogs; the "quasiliterary" dog to be found in comic strips and the like; the "quaint category" of talking dogs fabricated by the dog-food industry; and real dogs alleged to have the power of speech. Noting that Horace Walpole had used the word "dogmanity" to describe such allegedly "humanized creatures," Sebeok dismissed all the various historic reports—of Don, a conversational German setter, for instance—as illusory. Magician dog trainers give their animals secret cues, Sebeok said, and Don was made to seem a talking animal by suggestibility in the audience—his barks could be heard as merely barks when listened to independently of the suggestion that the dog was speaking.

As for Arli, "Borgese's style of writing is so effusive," Sebeok assured the readers of *Animals* magazine, "that her interpretations of her dog's feats are, to put it in the kindest way, ambiguous." Short shrift, indeed. "Our brief but close encounters with canid communications of the third kind have once again underscored the common error," Sebeok asserted. "The secret is concealed not in the dog, but in the man."

Or in the woman. As we shall see, it is a point that will be raised against nearly all innovators in the field of interspecies communication—*we imagine it is the animals who speak but it is really just the humans, fooling us, perhaps fooling themselves.* Sebeok is often the one who raises this argument, but there are many others who believe as he does.

It seems likely now that Arli the poet is a dog who existed mainly in Elisabeth Borgese's desire to find meanings beyond the meanings. But Arli the busy typist, who could *sometimes* make a connection between the spoken word and a snarl of symbols on his special typewriter—this Arli cannot be dismissed so easily. It may be that he was not merely the product of wishful thinking.

A DOG IN HISTORY: ARLI IN RETROSPECT

Other dogs don't type. His household companions, given the same initial opportunity as Arli, didn't learn to type. So Arli has that much claim to stand alone—a dog in history.

The motherly kind of pride that Mrs. Borgese felt toward Arli is clear enough, but the effort to push aside her results as not meaningful seems unfair. Except in the effort to find poetry from Arli's "cad-a-baf" days, Mrs. Borgese is a much more likely daughter for Thomas Mann than Sebeok's brush-off would indicate. Her book on the language barrier builds carefully on the work of Pavlov and others. It contains some truly baffling tales about elephants in Southeast Asia, and those inclined to think of her as jumping to conclusions might consider whether someone who will trek to India to gain firsthand evidence is taking the easy way out in seeking conclusions. Mrs. Borgese tried an extension of the Arli experiments on a young elephant and on a chimpanzee, which indicates, at the very least, an interest in cross-checking. Droll stories of women—and men—who are credulous about their pets are so commonplace that it may be tempting to dismiss Mrs. Borgese as one of these. On balance, she seems closer to Rensch and Pavlov than to a flutterhead. The general feeling that comes from the Arli account is that of a slightly scatterbrained dog with an ability to express occasional pure hunches and of a canny woman who knew the difference between proof positive and suggestive incidents.

What had really been achieved and what hadn't been achieved?

Leading a dog, by whatever means, to recognize words, numbers, and an abbreviated alphabet is a considerable feat in itself. It is not always clear, though, if this particular talking dog had any true grasp of what it was talking about. Sometimes Arli seemed to—but the evidence slipped away again for he was inconsistent, and even the best of his typing leaves trails of doubt.

At one point, Mr. Borgese discusses the difficulties that arose when sounds originally learned as whole words—a sound like *dog,* for instance—were broken into their separate letters. Arli, she said, was apt to get the right letters but in the wrong order. *Dog* spelled backward is *God*—that's one of the mistakes Arli made. Obviously, there are considerable language deficiencies if the dictator says *dog* and the typist responds *God.* (Even if the dog had a high opinion of his own breed, when you're taking dictation it's no time to be flip.) Although she calls attention to this problem and describes techniques for making Arli more precise, Mrs. Borgese does not really offer the conclusion that this kind of confusion was overcome. It leaves us with the feeling that Arli, who appeared to know who a *dog* was, had no knowledge that *God* was any different. The degree of Arli's comprehension of the meanings contained in the letters he struck remains a question mark.

To her credit, Mrs. Borgese knew this. She speaks, in many places, about the uncertainty of the deductions to be drawn from the typescripts.

"Dogmanity" is not that bad a concept, and may have benefited a bit from Mrs. Borgese's investigations. She had an insight that could be useful to others: that it is not impossible to use your own pets to look for the scientific and the profound. If she did not put her finger—or Arli's snout—on discoveries that are unchallengeable, she did raise questions about dogs' innate abilities with symbols and gives us hints that it may be greater than we had guessed. And that, after all, was her greatest aim—to get the question fully opened.

Although Mrs. Borgese's writings are not just an unstructured outpouring of wishful thinking, as Sebeok seems to suggest, neither are they proof of an arriving age of canine communication. It is not Mrs. Borgese's prose but Arli's typing that, in the long run, leaves us with an unresolved doubt. Much of the time, a long page of free-style typing by Arli resembles nothing so much as a typewriter operated by an automatic pilot that is on malfunction.

Mrs. Borgese's experiments with Arli were born only to be submerged by the growing excitement over the potential in dolphins and whales, chimpanzees and gorillas. But she had her own thought about a direction the chimpanzee experiments could take—as we shall see in the next chapter.

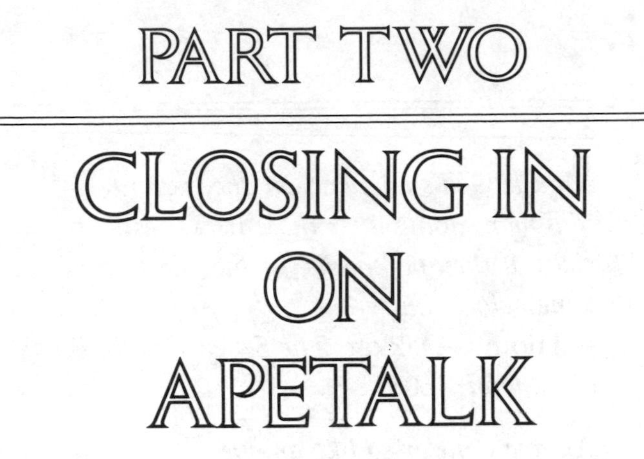

PART TWO

CLOSING IN ON APETALK

Man is God's ape, and an ape is zany to a man, doing over those tricks (especially if they be knavish) which he sees done before him.
—Thomas Dekker, *The Seven Deadly Sins of London*, 1606

An ape is ne'er so like an ape
As when he wears a doctor's cape.
—Thomas Fuller, *Gnomologia*, 1732

Acquisition of even the barest rudiments of language is quite beyond the capacities of an otherwise intelligent ape.
—Noam Chomsky, *Language and Mind*, 1972

Fine animal gorilla.
—Attributed to Koko the gorilla, 1978

6

The Upwardly Mobile Apes

A FRINGE BENEFIT OF THE SEARCH FOR APETALK HAS been that we see chimpanzees and gorillas more clearly now. It's their social status with humans that has been upwardly mobile.

The creatures that Dr. Lilly had placed at the fifth level of intelligence ("super animals," the anthropoid apes) should not be regarded as though their intelligence is the result of recent cultivation. We simply learned to recognize it and to teach the art of using symbols. This has now gone far enough that it may require a reevaluation of the difference between the human and the higher apes.

Two small vignettes offering glimpses of what it could be like to be a chimpanzee in the last couple of decades provide some notion of where—when they fall into the hands of humans who want to exploit their possibilities—these super animals have been heading. (The Lilly selection of the word "super" can seem strained, but judge whether it fits as you read these examples).

Dr. David Taylor, a veterinarian rather than an experimenter in animal language studies, specialized in exotic animals. He was ready to go anywhere in the world, on short notice, to treat a whale, a rhino, a gorilla, or the like. This gave him a ringside seat not only for observing captive animals in general, and the unusual lives they lead, but for witnessing the primitive forms of communication that have been used by humans to make those species deemed lower than ourselves "show their talent." He describes this in his book *Zoo Vet* (1977).

Often exposed to the seamy side of the zoos, Taylor had seen a keeper communicate with a bear by passing flaming newspapers under its belly to explain that it was time to go out and perform. He met a gorilla named Jo Jo who had grown sulky and selfish in captivity, so distrustful of humans that when Taylor put his eye to the spyhole into the gorilla's lair Jo Jo would send a whoosh of gorilla spit into the veterinarian's face. It was a primitive but meaningful way of saying, "Go chase yourself!" But it isn't a sentence and it doesn't have syntax.

Taylor's coolness in hectic situations often came to his aid. Once, while skindiving, he escaped the attack of a hammerhead shark by realizing that he must face it head on rather than turn away and kick for shore; panic communicates to a shark. There came a day when he needed that coolness just to have an exuberant time with a chimpanzee.

When a zoo in France had a problem with a bottle-nosed dolphin, the zoo's owner came in a car to pick up Taylor at his hotel. But wishing to show off a bit, he suggested that Taylor ride to the zoo on a motorcycle driven by the zoo's resident chauffeur, Henri, a chimpanzee. Taylor, who had a general distrust of European drivers and had never ridden with a chimpanzee, was nevertheless persuaded to climb aboard by the commissionaire of the hotel, who assured him, "Don't worry, monsieur. 'E 'as been 'ere to collect guests before."

The chimpanzee kicked up the engine. Taylor, clutching the animal for dear life from behind, scarcely believed he could be taking such a ride. A red light loomed. This was a perfect test, as it happened, for a question that was to become popular during the language experiments: *But do they really understand those symbols or is the teacher showing them what to pick?* Henri, the chimpanzee chauffeur, didn't pick a red light; it picked him. He knew what to do.

"Henri slowed down," wrote Taylor, "and, like the good motorcyclist he was, maintained our balance at very slow speed by weaving slowly from side to side until the lights changed. They did. *Brrrm-brrm-br-ooom!* Henri took us smartly away with a flick of the wrist and an effortless change of gear."

A bit later, the chimpanzee chauffeur had to swerve to miss a dog, and he was a bit over-flashy (lots of motorcyclists are) in the way he came to a stop in front of the ape house, but Taylor was alive and well. Not believable? Quarrel with the zoo vet. He was there. The chimpanzee cyclist was only one form of a more general phenomenon.

In their cages, in the communicators' laboratories, and sometimes in homes—where experimenters tried to give them a full enough life to motivate a desire to learn still more of human systems—a small handful of the higher apes have been pushing off, within the last twenty years, into new worlds. They aren't necessarily *better* worlds, but they are the worlds dreamed up for them as a way of determining just how far ape learning can go.

The seemingly irresistible movement into apetalk has old beginnings, but there is a feeling that it didn't become truly significant until the investigators were able to say, "An ape has learned to swear," "An ape has learned to lie," or went so far as to call a gorilla press conference. All of this is significant, certainly, and we will look at how it came about. Imagine, though, how differently certain of these chimpanzees must have felt from all chimpanzees, before or since. Here is a case in point.

One day, a chimpanzee named Bob, who had been torn from the mahogany trees of Sierra Leone, was transported to a house in Santa Barbara, California. This was certainly comfortable, but quite a change. Then Bob had to leave Santa Barbara in search of higher education. He was being sent to the Yerkes Regional Primate Institute in Atlanta so he would have the chance to really *be* somebody in the world. It is a sort of Charles Dickens story in twentieth-century form. Being a chimpanzee, Bob didn't have much, not even the clothes on his back, but he did have one treasure to take with him as he set out on this journey: he had his typewriter.

It was Mrs. Borgese who was seeing him off. The typewriter had belonged to the typewriting dog, Arli; it was a hand-me-down, which fits the Dickensian mood of this story. Mrs. Borgese is the only one, so far, who has believed that the way to take a chimpanzee into linguistics is to put him on a typewriter the way that chimp in France was put on a motorcycle. God, how hopeful it sounds! It's a wonderfully evocative happening, just what you might expect of an age of apetalk, but then a gloomy note takes hold.

What had *become* of Bob? Nowhere in the literature on talking chimpanzees could I find a trace of him after he left Mrs. Borgese.

Calling the Yerkes Regional Primate Research Center, in Atlanta, I began a small inquiry. Did they, perhaps, remember a chimpanzee named Bob? Several quick talks stirred a vague recollection. Yes, there *was* an ape with a typewriter—wasn't it an orangutan? Fifteen years had passed, many apes had come and gone, and it was hard to sort it all out. Finally I found my way to a former assistant director of the institute, General George Duncan, whose memory went back far enough.

"A typewriter? Oh, I think we did have that," General Duncan said. "It was in a bunch of junk down there and may have been thrown away." As for Bob, the chimpanzee, General Duncan couldn't quite recall him. Only the current superintendent of the institute, Jimmy Roberts, was likely to remember, the general said.

This was perturbing. You wouldn't want the very first chimpanzee who ever set off into the world of higher education with a typewriter to end up on the rocks, totally forgotten. The fear, as it happens, was baseless. Bob had not grown up to be a chimpanzee celebrity like Washoe, Lana, and other star performers of the primate institutes. But Jimmy Roberts certainly remembered him.

"Why, he's had a very good life," Roberts said. "He's still in our chimpanzee colony. Mrs. Borgese called him Bobby; we call him Dobbs. Dobbs has done fine, thoroughly enjoyed himself. It was quite a change when he came here from California. He had his own room in California, and his own TV. This was different for him, but Dobbs adjusted very well."

I felt a lot better, although I wondered why the typewriter hadn't come into actual use. There was an explanation for that. Something better than a typewriter—something simpler from a chimpanzee's standpoint and more sophisticated from the language investigator's standpoint—had provided a clever way to pursue the language studies. We will come to this in time—the system they worked out at the Yerkes institute that led to a chimpanzee language called Yerkish that was understandable to humans.

Other ape experimenters are building novel equivalents to Yerkish, and all this had to start somewhere. It certainly didn't start with Bob and his typewriter, nor even with the Yerkes center, although its founder began his work with apes early in the century. The ascent of the apes toward human-related language can be told quite briefly. It doesn't have that many chapters, but it goes back a bit farther than is usually recognized.

AN EARLY INTERSPECIESIST

The flareups over apetalk, though they became a particular passion mostly in our own time, had a brief and largely forgotten preview in Paris some two and a half centuries ago. Intellectually, Paris was the center of the earth; Parisians believed this devoutly, and probably felt they had seen everything there was to see.

Paris was startled, nonetheless, by the emergence of Julien Offrey de Lamettrie of St.-Malo, a philosopher-physician who disclosed that he had been trying to teach apes the art of speech. He jumped into all the contemporary Parisian squabbles with a vengeance. Both hooted at and hooting, for he was one of the great scientific disrupters of all ages, he was considered by his enemies to be a public outrage and a saboteur of the destiny of the human race.

"What were men like before the invention of language?" Lamettrie demanded. "They were beasts among beasts, at the mercy of their instincts." One invention only had been required, he claimed, to cause that beast to turn into the master of all others: language. Therefore, turn your attention to the education of animals, he said; present them with language, and apes can be raised to the level of the human.

His conviction that animals might be taught language if a proper method could be found was based on the body language of animals, which was there for all to see. Take the dog, he suggested: Did it not grovel in acknowledgment of guilt and was the meaning not clear to its master? Others have seen this, too. Didn't the dog show sorrow, gratitude, and bounding joy, in such measure that no human would mistake their meaning?

"It is a mistake," argued Julien de Lamettrie, "to suppose that human beings may be distinguished from animals through the operation of a natural law that enables the former to tell the differences between good and evil. For such a law operates in the case of animals, too."

If, at that moment, the apes he was training had spoken up in confirmation, Lamettrie would be saluted as one of the wisest of reasoners in that celebrated "Age of Reason." But the apes did no such thing. His experiment failed; the apes could not talk, and certainly not in French. Lamettrie removed the desire for interspecies communication from its reign in myth and legend to the very practical arena of the science labs. But it is primarily his boldness in asserting the pre-eminence of language as the force that gave humankind its power over the beasts that led to his long-term reputation as one of the precursors of evolutionary theory.

CHIMPANZEES' PROGRESS

Dr. Robert Mearns Yerkes, a comparative psychologist who entered Harvard in 1897 and retired from the Yale faculty in 1944, built a reputation as an adventurous primatologist connected to the most famous chimpanzee experiments in the first half of this century. These were usually laboratory experiments, but Yerkes became entranced when he watched the inventiveness of five young chimpanzees he set loose on a farm in New Hampshire. In birches and poplars, a chimpanzee named Prince Chim created a home that, to Yerkes, looked like a large bird's nest. It was revealing to note how easily the chimpanzee dealt with the need for shelter—humans, after all, find it hard to make do if left in a pasture to create their own homes.

Watching the dominance behavior of chimpanzees, Yerkes recognized that these animals have a complex social life with extensive interchanges between them. Describing his decades-long search to establish the dimensions of the chimpanzee intelligence in his 1943 book, *Chimpanzees: A Laboratory Colony*, Yerkes called it "inexcusable" to think of "the great apes as creatures of brawn or physical prowess rather than of intelligence." Trying to discriminate between types of intelligence, it was characteristic of him that he declined to say that chimpanzees are superior to ants in their way of life—for ants have all the benefits of regimentation and perfect organization—but he came to see the chimpanzees as individualistic thinkers. Ants sacrificed for the well-being of the group; a chimpanzee was highly social, using the social order as a way of expressing its own individuality.

That was an advanced idea. Yerkes remained at the top of his field, much admired and much imitated by new generations of primatologists, but he was always in defiance of the tradition that animals have no psyches and are machines, and that only humans are persons. To Yerkes, the anthropoid apes he came in contact with offered a great range of personalities. Prince Chim was a particular favorite—clever enough, it appeared, that he might achieve language itself. Yerkes' efforts along this line failed, however. Chim grunted and heaved in the effort to get out the single human word that meant most to him, *banana*. He assuredly knew what it meant, but Yerkes' effort to produce the word from Chim's throat went aground. He concluded that, as communicative as they were in many ways, the chimps would never master human speech.

He considered it evident, though, that chimps had a "simple language" for speaking to each other, and he produced startling examples of the extent to which a bright chimpanzee could understand humans and even, by a well-timed gesture, get a point across to these creatures who didn't know chimp talk. When a chimpanzee named Moos seemed out of sorts, a staff member inspected the animal's teeth and gums to see if they were the trouble. The dental inspection produced no clue, but as the inspector was leaving, Yerkes reported in *Chimpanzees*, "Moos took hold of his coat, drew him back, and raising his upper lip with one hand pointed with a finger of the other hand to a spot on his upper jaw." Moos's own diagnosis was correct—that's where the trouble was. The examiner "felt somewhat chagrined at having to be assisted in diagnosis by the animal himself."

Despite his own failures in teaching language, Yerkes had set the movement toward apetalk in motion, and it has never really stopped. Experiments in the Yale primate labs, during and after his own tenure there, would soon establish that chimpanzees can take symbol trips that indicate a high degree of creative imagination. One of Yerkes's associates, Blanche W. Learned, systematically tracked down that "simple vocabulary" chimps use with each other. She believed she could identify thirty-two highly specific sounds that conveyed particular meanings, mostly connected to what the chimpanzee liked to eat. While Yerkes and staff continued their explorations, others began far more elaborate experiments than the vocal training Yerkes had tried with Chim.

In 1931, a chimpanzee named Gua was raised right alongside a boy named Donald Kellogg in the home of W. N. and L. A. Kellogg, Donald's parents. What Donald received, Gua received. Gua and Donald were bathed, diapered, fed, and coached in spoken English as identically as the Kelloggs could manage to do so. The Kelloggs later wrote a book called *The Ape and the Child*, reporting that Donald grew up into a boy with a boy's habits and Gua grew up into a chimpanzee who couldn't talk. She had an understanding, they believed, of seventy words, but she never spoke.

When Carl Sagan was coming into prominence in the 1970s as a deft explainer of scientific ideas to the public, he described the Kellogg effort and others that followed it as the "twin cribs, twin bassinets, twin high chairs, twin potties, twin diaper pails, twin babypowder cans" method. It seemed to prove the reverse of the interspecies theorem.

"At the end of three years," Sagan wrote in *The Dragons of Eden*, "the young chimp had, of course, far outstripped the young human in manual dexterity, running, leaping, climbing, and other motor skills. But while the child was happily babbling away, the chimp could say only, and with enormous difficulty, 'Mama,' 'Papa,' and 'cup.' From this, it was widely concluded in language, reasoning, and other higher mental functions, chimpanzees were only minimally competent: 'Beasts abstract not.' "

BEYOND THE TWIN-BASSINET METHOD

Nobody immediately following the Kelloggs proved that chimpanzees could master people-talk, but in the 1940s philologist George Schwidetzsky reported that he had been studying chimpanzee talk and had gotten at least this far: he knew how to say hello. At the London Zoo Schwidetzsky gave a demonstration to prove his point.

Into a big cage of chimps, he said "Hello" in English. It produced absolutely no reaction, and Schwidetzsky walked away.

Then he strolled back, as though he were someone else, and said "Hello" in what he claimed was chimpanzee talk. The chimpanzees burst for the railing to be scratched and tickled by this stranger who knew how to greet them. Although he pointed out odd parallels between the style of chimpanzee talk and the tongue-clicking speaking style of the Bushmen, Schwidetzsky was no wholehearted apostle of the proposition that we might soon be conversing with chimpanzees. Chimpanzees expressed all sorts of emotions, he stated, but that should not be mistaken for sentence making. Attempts to establish that chimpanzees can "make a sentence" had by this time become the focus of those who devised language lessons for them.

Hans Kummer of the University of Zurich, in the comparatively recent (1971) appraisal, *Primate Societies*, contends that an ape's communication system has a primary limitation. Apes cannot express eventualities beyond "the Here and Now"; that is, they can make sounds to indicate their present desires, but can't make a date for Saturday night.

"The monkey," wrote Kummer, describing what he considered to be a fundamental gap in the communication system of all such primates, "can indicate that he claims access to a stand of mushrooms which both he and his partner can see at the moment, but he can communicate nothing about the location of the same stand when it is behind the next ridge, except by leading the other to the spot."

That may have been true of monkeys, but the notion that it was also true of their higher cousins was under challenge. Great abstracts like The Future, The Past, or Wouldn't It Be Great to Make a Million? had supposedly been too difficult for apes to reason about in a meaningful way only because of their lack of language.

Although the debate continues about an ape's ability to see past and future as a human might envision them, the experiments touched off by Yerkes' earlier findings were about to produce a chimpanzee commotion that would greatly enlarge ideas of what the higher apes can be led to think about.

Strange things were happening with chimpanzees in the 1940s—and they were happening at Yale. At the Yale Laboratories of Primate Biology, which Yerkes had founded in 1925 with the Rockefeller Foundation providing an assist, Dr. John Wolfe plunged ahead with experiments that are at least as provocative and revealing as the more recent run of sign-language and computerized investigations.

Wolfe's experiments hold many implications for communications research but, among other things, he proved that it is not solely the Here and Now that concerns a chimpanzee. Researchers knew that if a three-year-old child was put at the same pegboard with a three-year-old chimpanzee (the twin-bassinet method), the chimp would at first win at the game of putting pegs in holes, but the child would later pull ahead (remember the hare and the turtle?). The chimp's problem is that he can't keep his mind on the puzzle because his attention span is so short. There seemed no way to alter this fact.

To Wolfe must go a special credit for Chimpanzees' Progress. He took the familiar idea of a reward but made the winning of rewards into such a bucko romp for the chimps that they were soon revealing a hitherto-hidden power of concentration. To overcome the chimpanzees' tendency to forget what they were doing, Wolfe found a simple but wonderful device: the slot machine.

ADDICTS AT THE SLOTS

Writer Vance Packard, in the years before he turned his attention to operations of humans in the corporate structure, traced a long line of chimpanzee experimentation. His account in *Animal I.Q.* shows how quickly the human idea of "the intelligent animal" changes. Whales and porpoises had no place at all in *Animal I.Q.* Today, thirty years after Packard's study, they often dominate discussions on the same subject. It wasn't that they were considered dumb, but there was simply little in the way of evidence as to their intelligence until *after* 1950.

The stars of Packard's book are definitely the chimpanzees, and the most rambunctious of all are the chimps at Yale who went on the slot-machine kick. These episodes, instructive enough, have a kind of fury to them. They demonstrate that, whether chimps ever imitate human wisdom or not, they are quite capable of committing human follies.

Dr. Wolfe's Chimp-O-Mat, the chimpanzee slot machine, paid off in grapes. Wolfe began with Moos, that smart chimpanzee who had diagnosed his own dental condition. He demonstrated to Moos that by putting a chip in a slot Moos could win a grape. Moos understood almost instantaneously. At first, the chips were within easy reach, but Wolfe kept devising complications, making it harder to get those valuable chips that were good for grapes. He took the chimpanzees farther and farther into the rewards and hazards of prospecting. Making it more and more an act of intellect for them to get the chips that would satisfy their urges, he soon had created money-hungry chimps. They were emulating the slot addicts of Reno and Las Vegas, stuffing chips into the Chimp-O-Mat in search of a score.

That was the beginning of it, and the chimps had to discover Wolfe's tricks. They were not fooled for long when he substituted slugs for some of the chips. The slugs looked like the real thing but didn't work. The apes tossed them aside in exasperation.

Once he had the fever going, Wolfe introduced the chimpanzees to a Work Machine. The Work Machine would provide a genuine white chip, good for one grape, but not without sweat—eighteen pounds of pressure were needed to work the handle. Moos became the J. P. Morgan of the Yale grape industry by working the handle 185 times in a ten-minute period. "He was in such a rush to earn money that he didn't even pick up each chip as he lifted the handle, but instead brushed it to the floor where a pile was growing," said Packard. He knew the chips were worth grapes and saved them.

Now, with great piles of chips, the chimpanzees became wary of one another—they had to guard the treasure. Chimpanzee capitalism had been invented right there on the Yale campus. It went on through many phases as other colored chips were added to the scheme. Scrambling for success in their new little cottage industry, the chimpanzees learned that a red chip would buy water, a yellow chip a piggyback ride (on Wolfe's back), a blue chip twice as much as a white chip. White chips began to be scorned—who wanted white when the blues were a higher denomination?

Packard followed this dilemma even to the chaotic near-final chapter when a chimpanzee named Bula demonstrated her awareness that the yellow chip—good for a ride on the trainer's shoulder—might hold salvation itself. Dr. Wolfe had drastically changed the workings of the experiment. Bula had blue chips in her hand and was inserting them in the machine—but, instead of getting grapes, she was suddenly confronted by a rat that Dr. Wolfe had turned loose with her.

Her reaction was a weird counterpart to the behavior of those who dive for their hoard of treasures when they think they have to skip the country. Bula scrambled away from the rat, grabbed for her pile of yellow chips, and, says Packard, "took a flying leap up onto the psychologist's shoulder," screaming for deliverance—and ready to pay for it with the correct yellow chip.

A great moment, really, from the standpoint of the interspecies communicator. *The human had spoken, through poker chips, and the animal understood. The animal had spoken, with a poker chip, and the human understood.* Everybody understood except the rat, who must have been highly confused.

Chimpanzees might wander away from a pegboard, but the slot-machine adventures, even their perils, held them spellbound. Moos, who had milked the Work Machine for all it was worth when he saw the wealth it could produce, was able to determine later that he needn't return to the machine until his supply of chips was dwindling.

Chips were power—they bought what the chimpanzees wanted. Once he was ready to work again, Moos demonstrated that he could think well beyond the Here and Now. He would always draw at least a hundred chips, enough to last for a while. The chimpanzees treated the chips just as we do—as money. Red, blue, and yellow had specific meanings for them. Stirred by need and greed (a bit of each was involved), they mastered a heavy load of symbolism and proved that the chimpanzee is not at all immune to get-rich-quick schemes.

That great symbol trip of the chimps conducted by Wolfe had provided more than enough data to confirm that, somewhere in its heart of hearts, the chimpanzee *can* master increasingly difficult systems when the stakes are high. If the lab-based experimenters fell well short of giving the chimpanzee such an extraordinary "reward" as letting it weave through city traffic on a motorcycle, they were nevertheless thinking bigger. And the "twin-bassinet method," which includes the substantial reward of allowing a chimp to live like one of the family, was not entirely a failure, at that. It had not produced human speech, but that wasn't the ape's fault—it was his throat's fault.

Surmounting the vocal barriers to communication gave rise to the next great series of advances. The subsequent experiments showed that apes learn much faster, and can master more complex operations, when they are living in an intimate, familylike relationship, or even if they're just coddled a bit. As we begin to look at the experimenters who made the principal advances in apetalk from the early 1950s onward, one striking fact becomes apparent. Not all, but nearly all, have been husband-wife teams. Those who did not work as a family team nevertheless provided a crucial familylike atmosphere around the apes who were expected to show skill in human traits.

7

Chimpanzees: Experiences in Three Languages

UNLIKE MAN, THEY [THE CHIMPANZEES] DO NOT KNOW that they will die. Can I tell the ape that it will die? Could I arrange procedures that would culminate in a knowledge of death? . . . Until I can suggest concrete steps in teaching the concept of death without fear, I have no intention of imparting the knowledge of mortality to the ape."

This passage is from "Language and Intelligence in Ape and Man," an article by David Premack that appeared in the *American Scientist*. Premack, whose work we will be looking at, is the inventor of a kind of symbolic language for chimpanzees. The article reflected a general awakening on the part of interspecies communicators to the larger implications of the language studies. Much of the public had had a sporty gee-whiz reaction to the apetalk lessons, but the concept that questions as large as the ape's knowledge of its mortality were the real (if far-off) subject of the interspecies quest put a different kind of energy and excitement into the search. Are we headed then, to a time when the ape will be asked such questions as *What do you worship? Who do you like? How do you feel about lying? Does love have any place in the chimpanzee's life?* Fair enough questions on all counts, but, as Premack himself asserted, no attempt to "evaluate continuity between man and other species" could go forward unless the language experiments were able to resolve the question of the animal's ability in the area of what Premack called "self-reflectiveness."

Self-reflectiveness, he suggested, must not be taken as "an exotic feature of human competence, something found only on rare occasions or only in specially trained members of the population. On the contrary, though the developmental psychologist is only now studying the emergence of this type of behavior, it is evident that the roots of self-reflection can be found in childhood. Young children not only produce sentences, they judge the grammaticality of the sentences they produce. . . . They not only act intentionally, they know they have intentions, and impute them to others. . . . They not only memorize information, they know they have a memory. What is the status of self-reflection in other species? The topic is virtually unexplored."

In the ape-language experiments, the question of self-reflection became a vital thread connecting the work of the various experimenters. Without self-reflection, there can be no conversation at a significant level, no true sign of animal consciousness, and no creative intelligence operating beyond reflexes or instinctual behavior.

In less than ten years, not one but three languages for chimpanzees were developed—American Sign Language (Ameslan, or ASL), adapted from the language of the deaf and mute; the computer-based, lexigram-centered Yerkish lingo; and Premack-talk, a language in which the "words" consist of colored pieces of plastic in shapes and forms that are unrecognizable to anyone not trained to use them. All three languages have been shown to work, and all have infinite potential. How far chimpanzees can go on that potential was the question confronting each team of experimenters. They all shared a common goal—namely, to turn a few test chimpanzees in the direction humans have taken. They didn't expect the apes to become human or to fully achieve human powers of language and communication.

Could chimpanzees make the turn? And would this be any simpler than changing the rivers of the African grasslands on their course? Let's see what sort of answers emerged.

THE HAYES: A SPECIAL CLUE IN A LANGUAGE THAT DIDN'T TAKE

Movie film makes it possible to go back over an experiment and notice things that may have been missed when the experiment was actually taking place. The apetalk experiments contain several instances where further analysis supplied new answers, sometimes in defense of the interspecies theorem, sometimes not. The most critical incident of this kind—the one that first set the sign-language experiments in motion—was an insight from an experiment that seemed, at the time, to be a failure.

In Florida, in the 1950s, as a book called *The Ape in Our House* describes, Keith and Virginia Hayes tried to push a chimpanzee named Vicki into speech by absorbing her into a lively home filled with human talk. They offered her special rewards if she showed any progress. She did—at the rate of one word or less per year. Vicki, as Columbia's Herbert Terrace was to describe it later, learned "some unnatural tricks in order to obtain a reward." The unnatural tricks were saying "momma," "poppa," "up," and "cup." After viewing the extensive films Keith Hayes had made, primatologists had no trouble agreeing that the years of constant instruction really demonstrated that chimpanzees couldn't master human speech and probably didn't have the mental equipment for it.

Robert Yerkes took an early look at the film and saw something different. Vicki was quite a gesturer, and sometimes she indicated with her hands what she was unable to express vocally. Nothing came of this observation, though, until the 1960s, when another husband-wife team, psychologists Beatrice and R. Allen Gardner, saw the Vicki film and concluded that it was the *vocal* part of the experiment that had gone wrong. When the sound was turned off, an expressive Vicki became perfectly understandable by means of many of her gestures.

The Gardners conceived the idea that led to the first successes with apetalk: give chimpanzees a way to sign with their hands and they can learn a language.

THE GARDNERS: WASHOE SHOWS THE WAY

Beatrice Gardner was perfectly prepared in the art of noticing the kind of detail that easily slips by even a trained scientific eye. She achieved an early reputation as an observer of the hunting techniques of jumping spiders. Her husband came from an accomplished family and had attained distinction by studying rats. While brother Herb Gardner, who wrote *A Thousand Clowns*, and another brother, who invented the Nebbish, worked on the country's funny side, R. Allen Gardner was onto something serious: the mind of the animal. The Gardners, Beatrice and Allen, shared more than marriage. They had jointly proclaimed, "Our religion is that there must be some overlap between man and the animals." They saw themselves as able to demonstrate this overlap, and they did.

The Gardners have been much praised for intuitively sensing that Ameslan (ASL) could be adapted to any Vicki-like creature adept at inventing hand gestures to indicate meanings. They had wanted to use speech and signs together, indicating to the chimpanzee they settled on—a female named Washoe—that one could mean the same as the other. That's a difficult deduction for an animal to make in the early stages of a learning experience, and the Gardners reluctantly rejected it in favor of a system that forced them to do their own communicating through signs only. In 1966, they began.

"If she should come to understand speech sooner or more easily than ASL," the Gardners reasoned, "then she might not pay sufficient attention to our gestures. Another alternative, that of speaking English among ourselves and signing to Washoe, was also rejected. We reasoned that this would make it seem that big chimps talk and only little chimps sign, which might give signing an undesirable social status."

That explanation appears in a firsthand account of their progress in a 1969 article in *Science* magazine entitled "Teaching Sign Language to a Chimpanzee: A Standardized System of Gestures Provides a Means of Two-Way Communication with a Chimpanzee." Quietly stated, intense in its details—the Gardners are expert expounders—the news of the happy landing had just about the same effect on primatology that the news of Lindbergh's landing in Paris had on aviation. The Gardners had taken the first long flight in the history of animal behaviorism into the pioneer territory of chimp education. Their approach was highly personalized (you need to be personal with chimps to talk with them) but well documented and therefore scientifically defendable.

Washoe had learned to talk—not perfectly, not fluently, not as a human talks, but there were occasional signs of special insight in the words she signed and the observations she made. Before, if apes indicated some object, it usually meant "Gimme" or "Don't gimme." One day, Washoe strolled into the bathroom in the Gardner home and signed "Toothbrush." She didn't mean "Gimme" and she didn't mean "Don't gimme" (Washoe wasn't fond of having her teeth brushed). What she meant was "I see a toothbrush." As prosaic as it might sound, that—for apes—was new.

Washoe would say, "Sweet go." She also said, "Go sweet." Both meant the same thing: "Take me to the raspberry bushes." An ape who can say "Sweet go" and get herself a trip to the raspberry bushes as a reward for how well she is doing with her English lessons is certainly in a different position than your ordinary ape. Until Washoe, a captive ape who had tasted raspberries and adored them would have been unable to act on its desire except by watching for a human with a bag of raspberries and attempting to communicate, by gesture, that it should be given some berries.

"Sweet go" was a high point—one of the indications that language could prove useful in some way to the captive chimpanzee. An event of greater consequence, at least in the mastery of grammar, occurred when Washoe was presented with a mug that had a doll in it and declared, according to her trainers, "Baby—in—my—drink." The sign for "baby" is made by mimicking the act of cradling—a forearm is placed in the crook of the other arm to indicate rockabye baby. The Gardners had adjusted standard practice in ASL to signs that Washoe could readily understand or that seemed to come naturally to her. To say "flower"—she often wanted to say she saw a flower—Washoe would touch the tip of an index finger to either nostril. (In standard ASL, the nostrils would be touched in succession with the tip of the hand.) To say "dog," Washoe would slap her thigh several times. To say "I"—an extremely important word among humans, and Washoe was the first animal on earth to learn it—she would do what we do: point a finger to her chest. To say "clothes," she would brush her fingertips down her chest. The sign for "clothes" was like an extension of the sign for "I," although Washoe had no reason, as humans sometimes do, to mix up the inner self with the way she dressed.

At the beginning, Washoe's tendency to make incidental gestures was useful—the Gardners called it visual babbling and felt it was like a baby saying "goo-goo." Baby talk is important for human babies—the goo-gooing leads to speech. Washoe's hand-babbling served the same purpose: it led her toward relating specific meanings to particular hand movements. Later, when she wanted to say "shoes" she would hit the floor or her shoes with both fists. More subtly, if she wanted to say she needed a cover at bedtime, she would draw one hand over the back of the other. Hand-babbling had led to clear, if picturesque, language.

Sometimes the Gardners shaped Washoe's hands into desired gestures. (Later ape experimenters preferred this method.) The Gardners could also catch signs on the wing, so to speak. When Washoe's babbling produced a hand gesture that was close to an ASL sign, her teachers would "explain" the meaning by their visual reaction. When Washoe spontaneously made a sign that was like the sign for "funny," everybody would laugh and laugh, smile and smile, as they repeated the sign. After these rounds of chortling, Washoe knew how to say "funny." One day when Washoe was riding on the shoulders of Roger Fouts, the Gardners' student and principal assistant, she urinated on him and then signed "funny." It may have been a great moment in interspecies communication, but it also confirmed the same old chimpanzee recklessness. Washoe was by no means growing up to be a model of deportment, language or no language.

From single signs, Washoe went on to combinations. Even though she disliked the scrubbing that came with it, she could nevertheless say, "Hurry gimme toothbrush," as though to get it out of the way. A famous supposed signpost from the chimpanzee communication lessons developed when Washoe, seeing both water and a bird, signed "water bird." Sometimes the argument about the chimpanzee's language ability has unfairly swirled around this single instance of a Washoe deduction. Some observers thought she had conceived of just what humans see in duck, swan, or goose—a waterbird. Critics attacked this assumption with particular fury, contending that it credits a chimpanzee with "creating a concept" when she had done no more than indicate two separate ideas: "I see the water" and "I see the bird." The argument itself indicates that Washoe's performance was at least forcing the critics to consider whether she could speak and how far she could go with speech—what her special insights might be.

Her insights were certainly growing. She could lead a human to the refrigerator with signs for "Open food drink." The phrase "Roger Washoe tickle" meant that Roger Fouts should tickle her. Her trainer (who was Dr. Fouts to most people, but Roger to Washoe) claimed: "We taught her the sign for 'dirty' to indicate feces, and now she uses it for people who don't do what she wants to do." That was another first scored by Washoe. All the apes making progress in language lessons have come to share her fascination with words related to bathroom habits.

Fame brought the same thing to Washoe as it has to many other prominent Americans: a lawsuit. Four years and 160 signs after the beginning of the project, Washoe left the University of Nevada with Roger Fouts to take up a new life at the Institute for Primate Studies in Norman, Oklahoma. One day Dr. Karl Pribram, a Stanford brain scientist with more than a peripheral connection to ape-language studies (he helped get Penny Patterson's gorilla experiment started at Stanford), visited Washoe in Oklahoma to see how she was progressing under Roger's tutelage.

"I wanted to see how she was doing," Dr. Pribram told me. "I didn't know she was dangerous, and no one told me. She bit me, and my finger had to be amputated. They tell me that right after it happened she was signing 'Sorry, sorry, sorry.' I didn't see that. I wasn't in any condition to be reading signs."

He told the story ruefully, almost humorously—but he sued the keepers for millions. Pribram was certainly the last person such an accident should have happened to, for he is a gentle man, eloquent on the need for greater rapport between humans and animals. At seminars bringing together theorists in the field before the accident, Pribram spoke on such subjects as "What Monkey and Ape Can Tell Me About Human Language."

Although it was surprising when it happened, such an incident had actually been foreshadowed by the Gardners in their very first discussion of the Washoe experiment—its opportunities and its perils. In that opening report they wrote, "Affectionate as chimpanzees are, they are still wild animals, and this is a serious disadvantage. Most psychologists are accustomed to working with animals that have been chosen, and sometimes bred, for docility and adaptability to laboratory procedures. . . . A full-grown specimen [of chimpanzee] is likely to weigh more than 120 pounds (55 kilograms) and is estimated to be from three to five times as strong as a man, pound for pound. Coupled with the wildness, this great strength presents serious difficulties for a procedure that requires interaction at close quarters with a free-living animal. We have always had to reckon with the likelihood that at some point Washoe's physical maturity will make this procedure prohibitively dangerous."

Washoe had learned human signs; it may be that humans have not done as well in learning chimpanzee signs.

"The chimpanzee becomes adolescent," said gorilla trainer Penny Patterson, trying to explain what had happened between Pribram and Washoe and how, with a knowledge of chimpanzee language, it could have been avoided. "So do humans. Humans become pretty obnoxious around ages thirteen, fourteen, and fifteen. I was *really* obnoxious myself, arguing with my parents that I could do what I wanted to, that *they* couldn't tell me what I wanted. What happened with Washoe was partly the fault of the people there. They didn't warn Karl that what he was doing was not wise. . . . Karl was extending his arm to Roger Fouts to get a piece of fruit for Washoe and Washoe could see that as a threat. A threat by this new, suspicious person to her friend and parent-figure, Roger. . . . Washoe was reading the signs correctly from a chimpanzee standpoint, the standpoint it would have in the wilds. A chimp does not rapidly approach another, extending arms, unless something is happening, usually aggression. So there are rules. My guess is: Washoe was defending Fouts."

For those who didn't much care for chimpanzee experiments in general, the incident was a cause for snickering: *You can take a chimpanzee out of the jungle but you can't take the jungle out of the chimpanzee.* Perhaps this is true. Would a knowledge of language change anything? Would it alter the emotional balance of an animal who is often said to become dangerous—potentially, and sometimes actually, cannibalistic—after its early years? The old rule of thumb had been that a maturing chimpanzee becomes dangerous at age seven. Although the truth seems more ambiguous than that, and more a matter of an animal's individual character, no primatologist who expects to work with chimpanzees unscathed can afford to discount their potential for violence. Washoe had run true to form. She talked, but she also bit. Was it desirable to change this? Or possible? Can humans—violent creatures themselves—affect any change?

Dr. Lilly, in the very act of moving toward his definition of apes as "super animals" (he meant it in a restricted sense, not in a "superman" sense), had also placed them three full notches below the level of civilized man. His summation on the great apes was devastating, and certainly would be wildly disputed by the ape communicators. "The lowest-grade human moron," said Dr. Lilly in *Man and Dolphin,* with his customary fondness for stating a point at its extreme, "is above the highest genius in the gorilla or chimpanzee clan. One must look to the human idiots and imbeciles to find behavior corresponding to that of the primates. The tragedy of the radically 'mentally retarded,' as we call them today, is that they are not quite human and yet they are more intelligent than any of the primates except man." Such theories reflected ironically on the fast-breaking career of Washoe, the talking, inelegant chimp.

Dr. Lilly had another theory, later repeated and elaborated by Carl Sagan. Lilly speculated that humans, great aggressors in their own right, had in their rise to power killed off those "proto-humans" who were next to them in intelligence and had thus created a large gap between humans and their nearest competitors for world dominance. Sagan suggested that, because of our primordial crime in killing off the proto-humans, we owe it to the apes to do a rousing good job of bringing the higher primates forward into language. In the Sagan theory, the language lessons given to Washoe and others are a sign that the human race, after many millennia during which it has never atoned for its guilt, is now finally signing "Sorry, sorry, sorry" about the way we came up in the world, smashing rival species to smithereens.

What a great load of theory this is to rest on the gestural successes of a little chimpanzee! There could be some shade of truth in it, though, and whether it fits the evolutionary facts or not, there is no doubt that many humans who respect and are trying to teach animals certainly do so in a spirit of atonement for the lasting perfidies of the human race.

Nor was the effort to make atonement through language lessons in vain. In the hands of various educators, chimpanzees continued to make considerable progress.

THE RUMBAUGHS: TAKING CHIMPANZEES TO THE COMPUTER

The world began to learn that the talking-writing chimpanzees were, among other things, creative cussers. They used bathroom words to exhibit, at some low but definite level, creativeness in language (very much like the human race?). Washoe, whose vocabulary was eventually estimated at 160 words, had been quoted as saying, "Dirty Jack, gimme drink," but Lana, student of Duane Rumbaugh, Sue Savage-Rumbaugh, Timothy Gill, and others in the Yerkes Institute teams in Atlanta, was blunter. Sagan quoted Lana as saying, "You green shit."

The Rumbaughs and Gill, using Yerkish, elicited many creative expressions—not all of them in the form of chimpanzee cussing—from a series of bright chimpanzees. The swearing, though, was characteristic. It seemed to hold attraction for all the talking apes. Even Nim, the chimpanzee whose failures in language learning were eventually used in an effort to show that chimpanzees can't make sentences in the human sense, was assuredly a creative cusser. In *Psychology Today* Herbert Terrace, who conducted his own independent experiment at Columbia University teaching sign language to Nim, reported, "On many occasions, Nim signed *dirty* right after he had been taken to the bathroom. At other times he signed it and then showed no interest in using the toilet. The misuse of *dirty* was often accompanied by a slight grin and an avoidance of eye contact. Typically, it occurred when Nim did not want to cooperate with his teacher or was given a new teacher. Like a child saying, 'I need to go to the bathroom' when he or she doesn't have to go, Nim used the sign *dirty* to manipulate his teachers' behavior. Other apes have also learned to use the sign *dirty* to express a need to use the toilet."

Sign-language-using gorillas in the Patterson experiments showed similar traits. Before the gorillas, however, nearly all the imaginative verbal utterances from apes were the result of experiments managed by the Rumbaughs and Gill. Nim was the minor example in the field; Lana, star of the Yerkish lessons, was the major one.

What *is* Yerkish? It's a language that has the advantages Mrs. Borgese was aiming for when she put a dog and then a chimpanzee on a specially designed typewriter, but it avoids the technical difficulties involved in operating a typewriter. It's tied in to a computer for a projection of the results; the sheet of paper in a typewriter is replaced by a display console. The keys the chimpanzee presses, alone or in sequence, present lexigrams, each represented by one of seven colors and one of nine geometrical abstracts. The system is somewhere between the deck of oversized cards that Rensch's elephants used and the mishmash typewriter keyboard (numbers, letters, words, symbols) Mrs. Borgese used with her dog Arli.

Although in time he attacked aspects of all the language experiments, including his own, Herb Terrace conceded that the Rumbaughs had found an instrument (The Computer That Knew Yerkish) that in effect legitimized findings by keeping wishful humans from influencing the chimpanzees' choice of symbols.

"At the heart of Rumbaugh's procedure," Terrace wrote in his book *Nim*, was this "infallible computer through which all communications with Lana flowed. The computer always responded instantly to Lana's requests. Because the impassive computer could react in only one manner, it was impossible for the computer to transmit covert 'clues' as to the correct response. When human trainers are used, it is difficult to safeguard against transmitting such clues unwittingly."

Ameslan, by contrast, is an impressionistic language in which a twinkle in the speaker's eye can be as important as the hand signals. An Ameslan teacher will interrupt a human student to say, "Watch the face, watch the face—you're watching the hands too much, and you can't get the meaning that way." Like very rich conversational English, Ameslan has innumerable subleties, and therefore ambiguities, built into the very system. Those who were attempting to teach language to chimpanzees found it difficult to prove their case to skeptics unless they could put the animals into a more straightforward system.

Yerkish, as a computer language, has the advantages (sacred to most academic researchers) of diminishing the ambiguity inherent in such experiments and of yielding quantitatively measurable results. Part of the acceptance of Yerkish-speaking chimps derives from this factor alone. The experiments have a sanitized quality, with very little depending on subjective deductions by the experimenter. Lana's lessons could be documentated in the kind of terms that behaviorists were familiar with and could accept.

At the same time it should be noted that while there are scientific advantages to the Rumbaughs' system, it also places an inordinately higher demand upon chimpanzees than upon human children in comparable learning situations. In order to talk, the human child gets all possible help from its parents. The Rumbaughs supplied affection and a range of experience for the chimpanzees, but—in order to satisfy scientists that nothing was subjective—the lessons proceeded under emotionally sterile conditions. Whether this will prove the most effective way of transmitting language to animals remains to be seen, but at least it is a way of minimizing the suspicion that "the explanation is in the human, not in the animal," as critics charge. The computer becomes a machine dealer in this poker game of symbols—it never deals from the bottom, it slips no aces, and the chimp is on its own.

What was Lana able to do with The Computer That Knew Yerkish?

By absorbing the lessons to a point where she understood *sequence,* Lana amassed a vast range of human expressions—many of which she had never heard before in quite the same order—to formulate in her brain and project to the display console. That's what her woeful plaint "Please, machine, tickle Lana" had been: a careful choice of sequences. Saying it when night and loneliness were upon her made it seem almost certain that this was no coincidence; it was from the heart. Most of Lana's computer-written statements were not from the heart but from the stomach, or prompted by an itch to get more games going. She would say: "Please, machine, give milk"; "Please, Beverly, move ball into room"; "Please, Tim, give ball"; "Please, visitor, groom Lana."

It wasn't an accident that all these requests were preceded by *please,* nor was Lana just more polite than other chimps. The lexigrams were set up to indicate that requests begin with "please" (parents teach their children the same thing in a less visual way). The other choices were not as obvious, and Lana had to be clear in her mind about what she wanted to say in order to avoid producing a garble.

Certainly these atypical constructions—critics called them "rote sentences"—had common ideas and a common construction; one idea was much like another and they were repeated often. Still, Lana's vocabulary and her sense of sequence grew, and there were surprises in store. The vocabulary passed a hundred words and was headed up. She had names for the people she met (Tim, Beverly, Bill, Visitor); for foods of different sorts (nut, raisin, cabbage, banana, bread, apple); for food in general (chow); for familiar objects (box, can, blanket, ball, shoe, feces); for special things happening (movie, television, music, door-open, window-open). She also had a way of describing an object when she could name it by color but didn't know what it was: the phrase *this-which-is.* It led to such roundabout declarations from Lana as, "Please, Tim, move this-which-is-yellow into room." (Humans also have expressions like that, and in fact we *must* have them or we wouldn't be able to cope with the unknown: "I don't care what you call that whatyamacallit, you turn the motor up and it shakes my whole damn house!") By having a way of describing *even the unknown,* Lana was moving toward one of the more complicated functions of language, something it is common by thought that animals could never possibly express. With The Computer That Knew Yerkish, Lana managed a lot of things.

Clearly, though, Lana could spend many years signing "Please, Beverly, give ball" or "Please, Visitor, window-open" and the case for apetalk would not be substantially advanced. Stock sentences or sentences that always follow one line of logic, and one line only, raise the suspicion that Lana is achieving a certain mechanical skill but not disclosing an inner consciousness. In other words, we might come closer to showing that *a chimpanzee can simulate a computer* than showing *a chimp does independent thinking*.

In fact, to prove language *comprehension*, it is necessary that the animal use the right words at the right time.

Think of how it happens with a child. A child beginning to put thoughts together will discover that words learned independently can be formed into new combinations, and this is the essence of "learning to talk." Give the child just a few words like *hit, wagon, pole, bird, stop, chase, dinner, eat* and the combinations are limitless, especially if a few grammatical tricks in the form of prepositions and conjunctions and the like are thrown in: "Chase bird," "Eat dinner," "Pole hit wagon," "Wagon hit pole." The possibilities are so tremendous that a child very naturally begins to play with them, sailing words around like model airplanes—there are beautiful flights and there are nosedives. If the circumstances are right, the child may someday burst into the house shouting, "I chased the bird with a wagon and hit a pole." Although we don't often think about it, the hairbreath way that the child seizes just the right words at just the right moment is quite possibly the most prodigious of all human accomplishments.

To judge expressiveness in language, we have to know that the speaker *intends* to say what he or she says. "Hit the pole" may be a good description of an outdoor smashup and "Eat dinner" may be a good description of what is desired next; but if the speaker finds it just as easy to say "Hit dinner" and "Eat the pole," it's not fair to suggest that the creature has learned to talk. The talk, while colorful, wouldn't represent what is going on in the chimpanzee's, or anybody else's, mind. Only when a statement represents a fair appraisal of the situation (or an attempt to conceal the situation) can we truly say the animal is "learning language."

Now let's imagine that a chimpanzee has just sat down to the dinner table when a second chimpanzee comes along and wipes dinner off the table with a swing of its arm. If the offended chimpanzee diner pipes up "Hit dinner!" *then* we might agree that it is displaying a command of language. But does this ever happen?

One day, trainer Timothy Gill was testing how Lana would react to an act of sabotage. He purposely garbled the sentence she was making by touching a control that fed a wrong word into the sequence she was creating. Looking around to see what had gone wrong, Lana spotted Gill and signed "Please, Tim, leave room."

The right words at the right time.

To go back to Premack's stipulation about whether the ape experiments can show self-reflectiveness: Lana not only knew what sequence to hit, she knew that she knew. When it didn't happen, she knew that something, or somebody, was meddling.

The incident proves a knowledge of sentence construction, and it also proves that the animal is not just babbling along, trying to strike combinations that please the trainer, but that it has a deep consciousness. *The animal not only thinks but knows that it is thinking.*

The Rumbaugh experiments were not the first to show this, but they were the first to accomplish it with so many safeguards that it was difficult for behaviorist critics to deny the validity of their results.

If it were suggested that talking through a computer is an odd way to go about it, the Rumbaughs had a ready answer: No odder than sewing on a sewing machine rather than by hand.

"Having designed Yerkish as well as the computerized correlational grammar that parses Yerkish sentences," Duane Rumbaugh and Timothy Gill wrote in *Science* magazine, "we could hardly be unaware of the many ways in which it differs from English. Though we nowhere 'equated' Yerkish with English sentences, we contend that Yerkish is a language easily discriminable as such. . . . Lana can learn to do some of the things with Yerkish that *Homo sapiens* does with his languages. After a mere six months of study . . . Lana could complete correct sentence beginnings and cancel ungrammatical ones. Since then Lana has demonstrated that she can do considerably more."

Duane Rumbaugh's talent for keeping all this in perspective was exhibited in a recent conversation I had with him. I raised a point made by the British psycholinguist Jean Aitchison in *The Articulate Mammal.* After appraising Lana, Washoe, and others, Aitchison concluded that chimpanzees *do* seem able to learn language but have to be prodded into it; it's not natural to them, and they feel no affection for it. I had noticed this myself. Apes seem to go to their language lessons in the same spirit as a small child, not destined to be Jascha Heifetz, goes to a violin lesson. Conceding this,

Duane Rumbaugh said, "It isn't like the chimpanzees insist on staying overtime to work on their language lessons."

He also conceded that these reluctant learners are not necessarily going to achieve the dream of so many by going on to teach these humanesque languages to each other. "I would doubt," Rumbaugh said, "that the older will try to teach the younger."

THE PREMACKS: A LANGUAGE IN PLASTICS —AND COSMIC QUESTIONS

If I told you that Webster's dictionary and the principles of grammar could be duplicated by using colored pieces of plastic in odd geometrical shapes, you might give me an argument. If I added, "Also, it works better Chinese style," you might abandon me for more sensible company.

Dr. David Premack, working with a chimp named Sarah at the University of California at Santa Barbara, nevertheless invented such a system. He had help from his wife, Ann James Premack, in reporting what is probably the most esoteric of the chimpanzee experiments. Together they explained how Sarah learned to manipulate a plastic language of about 130 words as well as a number of special grammarlike or mathlike ideas.

This is mysterious territory—a large scale attempt to convey language by a kind of symbol that isn't very familiar to humans. "Chinese fashion" means that thoughts were expressed by taking the pieces of plastic, which had metal backs, and lining them up on a magnetized board top to bottom, instead of right to left, in order to convey a message. This experiment completes the circle of three talking chimps—Washoe, Lana, Sarah—who communicated in three different languages.

While it would be quite possible, following the principles of Premack-language, to incorporate equivalents for all those tiny but important words in the English language such as "and," "but," "of," "to," and "from," that isn't what Premack did. Sarah used a language that may have been comprehensible to her, but it would be hard for us unless we just plain sat down and let Premack school us the way he schooled Sarah.

For example, see how long it takes you to decipher the intended thought from this statement given to Sarah: "Sarah insert apple pail banana dish." I didn't find this at all simple to translate and was surprised that Sarah was able to. It means: "Sarah, insert the apple in the pail and the banana in the dish" (and you will receive rewards and praise).

One reason a human puzzles over the message—and quite possibly for longer than Sarah did—is because it approaches the form of a mathematical equation about as closely as it does that of the normal human sentence. The word "insert," the operating principle of the command given Sarah, functions much like a mathematical symbol. It explains the operation intended—*if the apple fits, pail it.*

Premack's accounts of the extended transactions made possible by this system prompted Eugene Linden, author of *Apes, Men, and Language,* to conclude that "what is a medium of communication for her mentors is, from [Sarah's] point of view, a series of multiple-choice problems."

Sarah was kept in confinement, usually caged, not brought up with other chimps, and she depended on Premack and his associates not only for companionship but for what she learned of life and its demands. It could be demanding, indeed; through Sarah, the investigators proposed to discover the truth of a series of special propositions. Using plastic symbols arranged in the right order, Sarah learned to make many of the simple demands that chimpanzees in the other experiments had mastered ("Sarah want apple"). As the Premacks pointed out, "If a small triangular piece of blue plastic is consistently associated with an apple in the ape's daily experience, whenever the ape later wants some apple, it will put the blue plastic on its writing board and be given an apple." That explained the principle, but the investigators were shooting for much bigger game.

At this point, it seemed clear—to the Premacks—that chimpanzees do talk in at least a limited sense. How much of their inner reasoning, or inner concealments, can be glimpsed in the course of their talking? Would the chimpanzee understand the proposition "If . . . then"? ("If I get a candy bar, I'll give you a kiss.")

If a talking chimpanzee knows somebody is trying to talk it out of its food, will the chimp reveal the secret place where the food is concealed? If the chimp learns to identify the "bad" person who tries to make off with its food, will it then learn to clam up and *not* talk or give signs even though it has become proficient at this?

These and other hard questions were tried out on Sarah. They were certainly language experiments, but they were also, as Linden thought, multiple-choice questions similar to those the behaviorists had been using for decades but taking new and surprising forms because the ape was known to be a talker, someone who could make deductions or spill the beans if she talked too much.

"Writing" vertically with the plastic chips, the trainer would set up the idea of "if" and "then"—setting out an apple and a banana and suggesting *if apple, then chocolate* (choose the apple and you can have some chocolate, too). Sarah learned *if-then* by figuring out that by making the right choice she could get the extra reward. Jean Aitchison, the psycholinguist who agreed that apes could talk but wasn't sure they could talk in important ways or had what the psycholinguist defines as language, demurred: "Note, however, that the fact that Sarah can understand the logical notion *if-then* does not prove that she has language. The relationship of language to logic is still very unclear." But many others feel that language and logic are like a pair of pliars. If an abstract deduction can be made from "words," then language is at an advanced state: that's how the ape experimenters accepted the Premack findings even though they seemed to have roots in equations found in philosophical logic as well as in dictionaries.

What about the experiments with sinister strangers? This was really an exploration of a chimpanzee's psyche. Does she know enough to cover up when trouble is on the way? Premack created a "bad trainer"—a person dressed in a special costume so that Sarah would have no trouble recognizing him if he turned up again. A contrasting "good trainer" also played a role in the experiment. He exposed Sarah to a series of tests in which she was elaborately shown where some fine tidbits of food were hidden. The "good trainer" wouldn't steal the food if he found out where it was hidden, but the "bad trainer" would. Was Sarah willing to share the secret of the concealed tidbits regardless of the consequences?

After she had been tricked by the "bad trainer" a few times, she would throw things at him to keep him away, but she couldn't keep herself from giving slinky looks at the spot where the food was hidden—talking in spite of herself. (It's perfectly possible that a chimpanzee who knows how to shut up is harder to discover than one who can learn to talk.) This experiment was really a test of memory: Could she recall who the "bad trainer" was and what happened when a fellow like that came around? After a while, she remembered so well that she no longer troubled even to throw things; she would hunch in the back of her cage and not come forward if the "bad trainer" entered.

Language depends, very largely, on memory, sharp discriminations, logic, and imagination. Each comes into play, and the various talking apes, including Sarah, displayed these traits to different degrees. Even though the experimenters went to extraordinary lengths to devise clever language systems that would bring out these qualities, it was always possible to find them elsewhere in the apes' other doings.

I take it as significant that the most revealing experience involving Sarah wasn't the result of her daily lessons in Premack's plastic Chinese but occurred when she was settled down in front of a television set to watch a videotape of a show about wild orangutans. Sarah watched intently. When a young male orangutan was captured in a net—a far more profound sequence was on the TV screen than in her plastic chips—Sarah hooted and threw crumpled paper at the television set. That seemed like interspecies communication of an advanced kind—the right boos and heckles at the right time.

Eugene Linden, who made an exhaustive study of all the chimpanzee experimenters for his book-length study of the chimps' progress, eventually found himself thinking back to a magazine article he had once read about Elisabeth Borgese—all those fine tales she had told about a dog who knew his name and elephants who did the inexplicable. He just wished, Linden wrote, that Premack had even one story to indicate that Sarah, caught up in her plastic language lessons, ever *personalized* anything.

Sarah's reaction to the plight of the wild chimpanzee, at least, was extremely personal. Television was giving her a broader view than the plastic chips could, something large enough to really interest her. More satisfying rewards and more intriguing adventures may be what the experimenters will have to offer in order to obtain a genuine insight into what is in an animal's mind. A chimpanzee can't really be expected to do big-time thinking while its options are banana/apple/chocolate. Suppose the language lessons went a lot farther and those options became: return to Africa/take a ride on a spaceship/have your own office/kick the experimenter out of the house when you feel like it?

Although the experiments with Sarah were highly materialistic (this is the forte of the behaviorist, and Premack is an eminent behaviorist), when Premack turned from the experiments to their higher implications—or when his wife joined in the discussions—he could become as cosmic as any of the ape experimenters had ever been.

The Premacks knew that humans, because of their ability with language, had created rituals, myths, and religions by the thousands. Premack's statement that he was not going to allow an ape in his charge to know about death unless the ape had a way to cope with its fear of dying seems admirable but impractical. There is no way to avoid this when language arrives, because the great secrets are the ones that most interest the tribe. If we study primitive experience, we immediately see that humans in back-near-the-Stone-Age society mused more intensely, if not necessarily more effectively, on life's basic mysteries than modern humans do. Part of the function of civilization has been to blur the preoccupation with trying to explain what we are doing here.

Whether apes know about death or not, it is quite clear that there are whales and porpoises who seem to have a pretty good idea about it. Fishermen watched in wonder as a porpoise hit by a bullet from a yachtsman doing target practice was supported by others in the porpoise band until it died. Other whales also stay with dying members of their group, in an almost ritualistic death-watch fashion. And who can be sure that apes know nothing of death among their own?

We can speculate, but we really can't be certain. Premack himself can't be sure until he is able to put it into a question: "Where do chimpanzees end up, old girl? Have they told you?" We are surprisingly close to the time when some form of such a question could be understood, but we will have to make sure that the answer we get is a real one, not a babble-answer provided by maneuvers with computers or plastic forms. If apes are at all like humans, such questions will be taken seriously, the essence of what is important enough to be talked about.

Chimpanzee experiments, with their framework of "askable questions," may or may not have reached their zenith with the experiments just discussed. The Rumbaughs, the Premacks, Roger Fouts, and others have lines of inquiry to follow and may yet come up with larger answers.

In the meantime, the gorilla experiments are exploring territory the chimpanzee experiments never reached.

When language experimenters finally realized that a human might dare to take gorillas on a large-scale symbol trip (and a gorilla is not easy to take anywhere), the language experiments began to touch areas that even the most elaborate earlier experiments had left alone.

For more than a hundred years, long after they had fallen captive, gorillas have been mystery creatures, especially to those who know them best. Gorillas may live in zoos but their minds seem to live elsewhere. Sometimes a keeper could see this—catching strange glimpses of the inner gorilla that didn't fit with all that was known about these animals. Only as the heart, mind, and psyche of the gorilla were explored by the most adventurous scientific investigators of this era did the ape experiments make their greatest advance. These experiments, as remarkable an adventure and conceivably as great an act of scholarship as anything the twentieth century has produced, were carried out by those who did not claim extraordinary scholarship and who usually downplayed the daring of their acts. The story starts in jungles and zoos, but the trail inevitably leads to a trailer-house in the hills above Stanford University, where, exiles all, two gorillas and the crew of the Gorilla Foundation explore large questions, defy conventional wisdom, and provoke the faith that an age of interspecies communication is fully upon us.

8

Profile of the Gorilla: Conversations with the Universe

SIGNS HAVE EXISTED ALL ALONG THAT THE GORILLA has much in its heart and mind and even a way to express it. Whether the gorilla wishes to express this for humans is questionable. The gorilla lies low, much like those humans who do not have the time of day for another being until they are certain you are worthy of acceptance, affection, passionate friendship. When they want to, gorillas can express their feelings with a skill that makes other animals seem like emptyheads. Rarely do they choose to do this—with a human, at least—but those who have been loved by a gorilla know they have been loved.

That the quest for interspecies communication turned toward the gorilla, and *had* to turn there, shouldn't surprise us in the least. Its resemblances to humans, its position at the pinnacle of ape intelligence, its disposition to family groups as the social unit—all this pointed to the likelihood that the gorilla would prove to be more of a talker than any other land creature, including the chimps. Many had noticed this, but for many reasons no one followed it up. Dealing with a gorilla isn't easy. Just *obtaining* one isn't easy. While there were several decades of experience behind the chimpanzee language experiments, direct experimentation in persuading gorillas to talk was nil. Then, when the gorilla breakthrough came in the 1970s and progress in gorilla talk tumbled along at a rate unheard of in any other interspecies experiment ever, the reports were so bizarre in comparison to those of the more typical chimp talk ("Nim wants banana") that the gorilla tales were sometimes lumped with other credibility-straining reports of psychic tricksters or horses solving math problems.

Still, the new gorilla accomplishments were not an astounding departure from earlier behavioral experiments, which had shown that gorillas can learn at surprising speed but that they really don't care much and would prefer not to bother. *If* they cared—*when* they cared—gorillas performed wonders, and always had. Motivating them was something else again. Gorillas kept their own counsel and maintained their aloofness.

Ngagi, a male gorilla whom we will get to know in this chapter, apparently was perfectly well aware that at any time he could open the heavy door that kept him captive, but he never bothered to demonstrate this until the need arose. Mbongo, Ngagi's companion, had become ill and was taken away. When Ngagi heard Mbongo's distress cries, he calmly lifted the door, which opened from top to bottom like a garage door, picked up Mbongo, and let his gorilla friend "come home." Had they known that the gorillas had figured out the secret of the door, the keepers would have been alarmed, but Ngagi used his knowledge only in a crisis.

Possessed of a natural reserve mixed with a natural imperiousness, gorillas cannot be converted overnight into shrill, woofing, mindless beggars like so many captives in the zoo. Gorillas tend to retain their dignity even in captivity. When the gorilla speaks—in a gesture language developed far beyond that of other animals—the sudden release of feelings will not necessarily be directed at a human being or even another gorilla. That same gorilla who, in the jungle, did great rituals for members of the gorilla family will seem, in the zoo, to be speaking to a family that isn't there, as though it were having a conversation with the universe.

THE GORILLA SAYS NO WITH A QUILT

Europe's first living gorilla, a female, resided at the Berlin Zoological Garden, and she did not come cheap. They paid $4,000 for her when she was only three and a half feet tall. That was more than

$1,000 a foot, a most extraordinary price in 1876. The gorilla proved to be worth it.

She communicated. One of her clearest communications occurred when her keeper, who normally treated her with fruit, sandwiches, milk, and tea for supper, tried to treat a passing illness by feeding her quinine. From that time on, whenever that keeper approached her with a medicine bottle, she pulled a quilt over her head. *The animal signals and the human understands.*

Her habits were quite humanlike and, since she happened to come to Europe in the late nineteenth century when the struggle over Darwinism was both frantic and fanatic, it was a comfort to see such signs of gentility as the way she drank her morning milk from a glass and handled spoons with a certain adroitness. She knew soon enough which table manners were preferred, but if it seemed to her that no one was looking, it was spoons away as she hastily seized the dish and slurped its contents with no further attempt at politesse.

Juggling a tea cup full of hot broth, polishing off a hearty meal of chicken, rice, potatoes, and vegetables, going to sleep under a nice woolen quilt—why, there *was* something *hausfrau*-ish about her domesticated ways.

She must have dispelled some of the alarmist reactions to the arguments about Darwinism, for she did not at all fit the picture that anti-Darwinists had conjured of the monster ape who was said to constitute the link between animals and humans.

It is sometimes said that humans in the mass are *afraid* that we will give animals speech. I don't see any signs of such fear, but it is quite clear that humans in the main don't wish to feel a great rivalry with animals, either. Roger Fouts, who took over as Washoe's long-term teacher, has contended that Western man experiences a type of "divine trauma" whenever he is "forced to face the fact that he is not the center of the universe." To be on top of the ape heap is not the same as being the divine creation around whom all creatures and universes wheel. To be the *best* talker in these earthly climes is not the same as being the *only* talker—the single fount of creative wisdom. Darwinian theory suggested that humans are a stage of life rather than its brightly burning, solely imaginative core. Humanity shook down rather quickly into those who (*a*) felt diminished or (*b*) were perfectly glad to have new relatives. Those who remain uneasy about a close evolutionary connection to the great apes are assuredly far less numerous than those who are good sports about it and who would invite the gorilla for tennis if they could. Still, there are those who have had a rather good reason for taking no special notice of gorilla talkativeness. A talking gorilla interferes with the neat solution certain scientists found when they restored the idea of human *centralness* by treating the power of speech as though this were the magic amulet that has *rocketed us to the center* when we were, before, merely animals. This is untidy stuff, slippery reasoning. It is also quite a real factor in the interspecies battles.

THE GORILLA SPEAKS – AND THE HUNTER REFUSES TO UNDERSTAND

Looking back, we can discover that gorillas were using powerful gesture language from the moment the first white hunters and exotic animal collectors arrived in the gorillas' natural territories. In the beginning the gorilla stalkers may genuinely not have understood the gestures when a gorilla zinged from the underbrush and loomed before them, giantlike, beating its chest, scary as some oversized medicine man. The *first* time they saw this display, the hunters could not know that it was a bluff, a ritualized attempt to deflect them. Later they did know. And then it was necessary to pretend that the gorilla's cry meant "I kill" instead of just "Go away." The hunters made their adventures seem as harrowing as possible so the killing of those largely peaceful vegetarians, the gorillas, would seem a respectable act. To this day, gorilla poachers act on the devil theory, describing the gorilla as a jungle madman who must be killed to protect the tribe from its bloodthirsty ways (and, incidentally, for the money that comes from gorilla meat or the gorilla ashtray made from a severed simian hand).

The first of the white gorilla hunters, Paul Belloni Du Chaillu, didn't care a fig for the meaning of a gorilla's gestures. He had shootings in mind, not dialogues. Du Chaillu was no more interested in knowing the gorilla's real intentions than the trout fisherman is in asking the fish why it flops. An explorer, he trafficked in the riproaring adventures that brought him fame and money. The better he listened to the gorilla, the less he had to gain in his career, so he was no listener. Cannibals and gorilla kills were the stuff of a Du Chaillu adventure. The remarkable "Cousin Paul," as he later styled himself in the adventure books he wrote for children, slaughtered one gesturing gorilla after another and was chestily proud of all his feats.

The description he gave of the very first gorilla to fall before the guns of the white man has been recited in virtually every gorilla book since. This gorilla along the Ogowe River in Gabon, West Africa, reported Du Chaillu, was a creature of "powerful fangs" and "thunderous roar . . . his eyes began to flash fiercer fire as we stood motionless on the defensive. . . . And now truly he reminded me of nothing but some hellish dream creature—a being of that hideous order, half-man, half-beast, which we find pictured by old artists in some representations of the infernal regions."

In bringing the gorilla to full public notice in America, the adventuring hunter provided everything a headline writer could want. He depicted the creature as a stormy, death-dealing, sex-mad destroyer, an ogler and raper of women. Such hairy oglers and rapists—inhabiting comfortably far-off lands—were just the thing to bring a delicious shiver to genteel souls nibbling toast at the breakfast table. Du Chaillu's timing could not have been better. In 1859, the year the hunter came back from the Gabon, Charles Darwin's *Origin of Species* was unleashed on the world, stimulating the public appetite for tales based on natural history. For Americans brooding over impending civil war, Du Chaillu's stories of harrowing adventure provided a form of exotic distraction.

Today many doubt the reality of gorilla talk. When Du Chaillu first returned from gorilla hunts along the Ogowe River, many in the general public doubted the existence of gorillas, considering them mythical beasts. Scientists and offended explorers claimed that Du Chaillu was a braggart and a liar. They granted that he had been on the West African coast, but they claimed that his maps contradicted the known facts about equatorial Africa and, besides, they were such a tangle that no one could follow them. Du Chaillu had no photographs to prove his triumphs over the gorilla, and the skulls and skeletons he brought back could have been picked up along the coast, or even faked. The scientists thought they smelled doctored evidence. His specimens were snubbed by fledgling American museums that didn't quite see the use of them.

Outrageously pilloried but outrageously honored as well, Du Chaillu was the perfect nineteenth-century version of culture hero and animal naturalist. Du Chaillu came across as a king of derring-do, and many a scientific scalpel was poised to slit his overactive throat. He was, however, more than able to hold his own. A media star before radio and television were around to confuse the issue (although the journals and newspapers of the day did very well at confusing it all by themselves), Du Chaillu was a master at grabbing headlines, upside or downside, applauded or accused. He left for Britain in a huff, and did still better in the British Isles.

When Du Chaillu did not find things going his way at a meeting of the British Anthropological Society, he climbed over the members' chairs to shake his fist in the face of the chairman. An expert putdown artist named William Winwood Reade, a British writer, went off to Africa himself in the effort to prove that Du Chaillu was a fraud. In a condescending but effectively insulting statement, Reade declared, "M. de Chaillu has written much of the gorilla which is true, but which is not new, and much which is new but is very far from being true." Hadn't Du Chaillu blustered that before he shot them, the great male gorillas would "beat their chests like a drum"? No gorilla, Reade said flatly, had ever beat its chest like a drum.

Outraged by such assertions, Du Chaillu felt compelled to defend his honor by returning to the Ogowe for fresh evidence to use against his attackers. The scientists who opposed him chortled and wished him ill. Du Chaillu took a camera with him this time, and the pictures he brought back started an Ape Rush that has not yet ended. The hunters went, the collectors went, the tourists went. And no matter how much the apes retreated, they were unable to effectively communicate this message to the oncoming hordes: "You're crushing us. We must have breathing room."

MRS. BENCHLEY AND THE DRUMS

Of course, true naturalists who went to live with gorillas in the wild got the message; so did Belle Benchley, a forerunner of the gorilla communicators, who made it her business to poke into the gorilla psyche to find out what it had to say.

In all the annals of zoodom, there is no story quite like that of Belle Benchley. A sheriff's daughter and a lover of animals, she came to the San Diego Zoo in 1925 on two-week assignment as a bookkeeper. She rose to become its director and to leave her imprint on the profession of zookeeping in general. Her approach to zookeeping sprang from the same character traits that made her such an inspired student of the gorilla.

Mrs. Benchley believed that wild creatures would do better in captivity if some semblance of their wild homes were re-created in the zoos. More than this, she believed it could be done—would *have* to be done. She used every trait she had, including her own longevity, to implement her ideas in a lasting way. The San Diego Zoo's fame and distinction are a part of Belle Benchley's legacy.

The zoo collected primates. Belle collected primatologists. Robert Yerkes was widely known by the time he arrived to examine and study some of Belle's favorite creatures. Dr. Harold Bingham, the immediate predecessor of those like George Schaller and Dian Fossey who gained worldwide renown for their expeditions into gorilla country, conceived the notion that it was possible to conduct naturalist studies of the supposedly ferocious wild gorilla at close quarters in Africa. Bingham was not able to get as close as he desired, but he pointed the way toward studies that took on a great importance. Bingham, too, came to study in Belle's zoo, finding there what he had not been able to get close enough to in the African wilds.

Mrs. Benchley encountered puzzles in her investigations, and she liked nothing better than to divert these exquisitely curious, probing gentlemen to work with her own animals in her own zoo.

When she finally made a case for her own findings on gorilla language and gorilla character, by means of acute observations in her book *My Friends, the Apes*, published in 1942, a tone of exasperation crept in regarding the kind of scientist who hears little but judges much. Belle would listen, listen, listen to the gorillas. She would watch and watch. It is possible that no one has ever watched better. She did not try to decode gorilla language in a formal way—she was not like the chimpanzee experimenters we have looked at—but there was a great temptation to move from what she heard to what it meant.

The profile of a gorilla that she created still seems prescient and illuminating in every respect. Her judgments on gorilla character explain some of the oddest parts of the Patterson experiments with Koko and Michael (which we will discuss later), for they provide some clues on what the gorilla is like inside that remote, considering, imperious mind. Quite aside from what she had to say about gorilla talk, Belle's style of befriending animals in order to understand them has affected those she taught and has influenced the field of animal behaviorism generally.

Anybody at any age who was determined to commit his or her life to animals, feeling it the greatest profession in the world, became part of that clutch of people Belle Benchley carried along with her. Kenholm Stott began buzzing the San Diego Zoo in search of a job while he was still in grade school. Mrs. Benchley found him and gave him a job—at the age of eight. She didn't concern herself with child-labor laws.

Stott had read about the African explorers and, since he thought he could become one, he decided he might as well get started. He eventually became Mrs. Benchley's chief curator, making seventeen trips to Africa. Flat on his back when I spoke with him, Stott had been felled by a combination of African diseases that caught up with him after a lifetime of jungle exploration. He had no regrets and was still high on Mrs. Benchley. She had died seven years earlier, in 1973, very likely maintaining to the last what she had maintained all her life: "I am not a scientist." She could not do, she said, what scientists do; she did not have the training. She'd had no training for becoming director of a zoo, either, although it didn't stop her from influencing the whole profession. Stott seemed most cheerful, talking about her, when he recalled Belle's flutters of modesty on the day one of the most famous of the ape investigators arrived. When the visiting scientist asked Belle to accompany him on his visit with the animals and make notes, she protested how valueless they would be compared to his own. But the visitor insisted. So Belle took notes, and when they came back to the office the scientist sat down to see what Belle had written.

"He read them," said Kenholm Stott. "Then he took his own notes and tore them up and kept Mrs. Benchley's."

If she hadn't been a superb note taker, there would be no discussion here of Belle Benchley and we would know much less about the drumming habits of gorillas. The experiences she recorded with the gorillas Mbongo and Ngagi will explain the reaction of the scientist who threw his notes away. They also explain much more than we would otherwise know about the nature of an animal that nearly everyone, from Paul Du Chaillu onward, had greatly misunderstood.

The gorillas arrived at the San Diego Zoo in 1931. One of the great pursuers, Belle had acquired these animals in her usual manner. She wrote what she called "timid letters" to the African husband-and-wife exploring team, Martin and Osa Johnson, and pursued with such timid relentlessness that they couldn't deny her even though half the zoos in the country were angling vigorously for gorillas. Belle had exotic creatures in profusion, but the gorillas became her favorites.

Books on gorilla lore had led the zoo to expect creatures who would either remain imposingly silent or erupt into hellish roars and chest-beatings. Mbongo and Ngagi contradicted nearly all that the keepers and Mrs. Benchley had ever read. The gorillas' sounds came through closed lips but they came often, sometimes in volume, and they also communicated a great deal with their hands and bodies. And although they had endless ways of creating sound effects or signaling their feelings once they felt at home in the zoo, it was their drumming that most interested Belle Benchley. It was not limited to the chest-beating the books had mentioned, which represented only part of the gorilla's acoustical effects. Mbongo and Ngagi drummed on various parts of their bodies and on other resounding surfaces, and in doing so they produced so many different sounds and rhythms that Mrs. Benchley suspected the drumming wasn't just nervous sound-making but contained messages or some special meaning related to the gorillas' moods. She began to watch more closely, to speculate on what might be meant by various sounds, and to write her speculations down.

In time, her notes on what seemed to be a gorilla communication system of surprising variations and adroit game playing became the one real storehouse of information on everything in the behavior of captive gorillas that seemed to resemble "having a language."

An early, cryptic incident involving the gorillas and Osa Johnson stuck in Mrs. Benchley's mind. Osa, like Mrs. Benchley, was a miracleworker around animals, and at the time the Johnsons were public luminaries. Their African documentaries had opened in New York to huge premieres, just like the biggest Hollywood films, and their exploits had caused great excitement among schoolchildren, some of whom could hardly wait to invade jungle kingdoms on their own.

One day Osa Johnson came to see Mbongo and Ngagi, who greeted her with high-pitched notes, different in quality from any sound that Belle had ever heard from them. Osa answered with a whine and a half-whinny, and they fell together, then, like old, old friends.

EXULTANT IN THE RAIN

Gorilla alarm or mating calls are clear enough, but Belle also identified other sounds, including soft noises of comfort when one of the gorillas was distressed and the other gave solace and their sharing of "Hmmmmm!" "Hmmmmm!" sounds as they snuffled in their food and exchanged contentment. A strain of subtle chatter often seemed to be passing between them.

The drumming was more mysterious; the changes and breaks were so intricate that it was sometimes exhausting to follow sequences from one gorilla to the other.

It was not that Belle saw language itself in all this drumming; it was more like a reaching out by the gorilla to express what was inside him, thoughts that otherwise could not come out. The dancer does this and the drumming could have been, to some extent, improvised dance: toe-dance, foot-dance, paw-dance, slap-dance. Belle was aware that the transmission of information in Africa via drums can be amazingly fast. (It has been claimed that when Queen Victoria died, the news arrived first by the drums and then by the telegraph.) Belle did not confuse tribal drumming with the drumming of the gorillas, but she had a question about it:

Had the drumming spread from the gorillas to the Africans or from the Africans to the gorillas? No one could know—but she wondered.

Although the drumming remained mysterious, Mrs. Benchley often sensed the moods that it reflected. It could dominate the gorillas' lives, this drumming—fast patter; slower rhythms; the exultancy of chest-beating in the rain; the calling to each other through drummed invitations to come and play. There were games of foot-against-foot, when they drummed on each other, in elaborate patty-cake variations. There was gloomy drumming when a darkness seemed to fall across them.

George Schaller, who tracked gorillas on their native range in Africa, eventually offered an account of their drumming that resembles Mrs. Benchley's although he placed more emphasis on the patterning while Belle stressed the looseness, the unexpectedness, the way the drumming would vary and change as the gorillas seemed to find additional styles to express what was inside them.

The following is from Schaller's *The Year of the Gorilla* (1964), describing a confrontational chest-beating:

> The complete sequence, which is rarely given and then only by silverbacked males, consists of nine more or less distinct acts. . . . The females and youngsters in the group know that the hoots and the placing of a leaf between the lips are preliminaries to rather vigorous, even violent, actions on the part of the male, and they generally retreat to a safe distance. Just before the climax, the male rises on his short, bowed legs and with the same motion rips off some vegetation with his hand and throws it into the air. The climax consists of the chest beat, which is the part of the display most frequently seen and heard. The open, slightly cupped hands are slapped alternately some two to twenty times against the lower part of the chest at the rate of about ten beats a second. . . . Gorillas do not pound their chests with the fists, as is often stated, except on very rare occasions. Chest-beating is not at all stereotyped in its application, and the animal may slap its belly, the outside of its thigh, a branch, a tree trunk, or the back of another gorilla. One juvenile patted the top of its head about thirty times,

> and once a blackbacked male lay on his back with legs stretched skyward, beating the soles of his feet. . . . The grand finale of the display consists of a vigorous thump of the ground with the palm of the hand. The performer then settles back quietly, the display completed. It is a magnificent act, unrivaled among mammals.

Now we have Mrs. Benchley, in *My Friends, the Apes,* describing the evolution of Mbongo's and Ngagi's drumming as they were maturing:

> On rainy days, too, the beating upon the dripping chests was almost continuous. . . . Mbongo would stand up on the wide triangular shelves in the corner and stamp up and down, pounding his heels alternately hard and fast in almost the same rhythm as when producing with his hands. . . . But it is not always a happy story I have learned from the drums, for now and then, especially when they were younger, it seemed a little pathetic. . . . Sometimes as they beat they stopped suddenly and seemed to listen as though expecting an answer from some other youthful member of their band who might be feasting on tender shoots just beyond their sight, and often, as though disappointed, the youngster might stand and beat again and run off around the log, for surely just beyond he might find someone to join him in his play, someone he had lost. But the intervals between these solemn drummings became so long that eventually we . . . realized . . . that the world of our two gorillas was now encompassed only by each other, we few whom they had accepted, the familiar setting of their cage. . . . Huge and heavy as they were, the giant gorillas would stir into action late every afternoon, . . . tearing about on all fours with speed almost unbelievable. . . . They would stand erect now and then to issue the old challenge of the jungle, or running at full speed pause, one after the other, to beat loudly upon the resounding metal doors—the notice to their greatly beloved keeper to hasten for night was due [and] it was time to fill the huge stomachs with much food.

Mrs. Benchley's conviction that the drumming was full of clues to the inner gorilla led me, as it must have led many others, to consider all the rest of the gorilla's communication system. Was the system meant to communicate with other gorillas? With human keepers? With animals in the cages beyond? With all who live? With imagined figures in the sky? With all that is visible or invisible upon the earth—with the very spirit of the living world?

To Belle, it seemed increasingly superficial to dismiss the whole rich medley of animal sound as consisting of a few simple signals, some repeating songs and calls, and little more. In the song of the gibbon, which could be heard each morning, there was something stereotyped and repeating; it was different than birdsong but seemed to fill the same purpose. She loved the "high and sweet call of Negus, my lovely gray gibbon," but she made no case for the reality of a varied gibbon language. Gibbons had limits that she did not find in the gorillas.

How far did gorilla language actually go? Belle didn't know but she had her suspicions. Gorillaness, it appeared, was quite as complicated, in its own way, as humanness—complicated and full of expressive signs. "Communication between animals is quite evident," she asserted, "and sometimes I think almost as definitely understood between them as is human speech by men." She listened to human speech, including those languages that do not have vowel sounds, and concluded that there was a duplication of that richness in the "varied tones and significant inflections of many animal voices." Her words foreshadowed much stronger pronouncements from the contingents of interspecies communicators who soon followed her.

BELLE DESCRIBES "GORILLANESS"

In *My Friends, the Apes,* Mrs. Benchley provided the following comments on Mbongo and Ngagi, giving us a picture of their "gorillaness":

> To sum up the mind and character of a gorilla as revealed by my own purely personal study of our two, I should say they were sufficiently intelligent to learn to employ simple me-

> chanical things but were not essentially interested. They had excellent memories and they sized up a situation understandingly and rapidly before they even attempted a solution. They were impatient with futile effort, apparently lazy, and perhaps easily discouraged. They were inclined to follow the line of least resistance, although they were not cowardly. They were very much more creatures of routine, perhaps much better organized as individuals, than any other of the apes in our collection. They did not recognize either a stick or a rod as a tool but feared such things as a weapon and would cower back at a slight gesture. They were much more alert to sights and sounds than their apparent self-absorption would indicate.

Confronted by camels, Mbongo and Ngagi would start an outcry that put the whole zoo on its ear. When a visiting professor took a microphone away from Belle to address his students, Ngagi rushed to the bars of his cage, ready to tear them apart in order to return the mike to Belle. But in the face of anything the gorillas considered just normally dangerous, Belle found them "conferring with each other and agreeing that danger was near, and they would draw close together and by being very alert almost seem to protect each other in every direction from an approaching disaster."

Later, when serious efforts were begun to translate gorilla talk directly and to persuade gorillas to accept the human symbol system, the principal obstacles were those that Belle Benchley had identified with Mbongo and Ngagi. For all its understanding of method, the gorilla is recalcitrant. It can often work the scientist's puzzles with ease but may disdain to do so.

It is Belle's attempt to explain the mysteries, the powers, and those discordant notes that hold the gorilla back, which makes her seem so important to the interspecies quest today. She was more than just a fine woman who, for some reason, caused the most expert primatologists of the time to want to be in touch with her. By not thinking too weightily of how scientific everything was, she managed to focus on the inner workings of the animal brain. If what she suggested were hunches, nothing so far noted by more scientific observers has contradicted her. She called men like

Yerkes and Bingham "the real scientists" who would have to make the breakthroughs, but new discoveries in natural history are not made merely by pushing a graduate diploma in a slot as though it were a magic time card entitling bearer to One Discovery. Mrs. Benchley made many discoveries. The profile of the gorilla she developed is the deepest, most incisive guide to be found on what moves, motivates, and inspires the captive gorilla to share some part of its moody self with the human observer.

Belle Benchley lived to be ninety-one. If she had lived another few years, she would have made it to the late 1970s, when the language experiments with gorillas came fully alive. She might have had a real chat with a gorilla. As it was, the zoo she directed for so many years was full of attempts to communicate with gorillas, even if they usually stayed in the category of loving care rather than scientific experiment. One day in the 1970s, Steve Joines, a young primatologist-in-training, took on a Belle Benchley sort of assignment from the San Diego Zoo, which asked him if he could teach a gorilla how to be a mother.

THE GORILLA'S APOLOGY

The gorilla's name was Dolly. She had rejected her firstborn, removing it to an opposite corner of her cage and ignoring it. No sight on earth is any lonelier than that of a newborn gorilla whose parent sits in a far corner, tense with displeasure, showing not a sign in the world of acting like a mother.

Zoo authorities did not intend to be stymied or to let Dolly miss the joys of motherhood. They took Dolly's baby away for protection while they contemplated what deficiency there might be in the gorilla's training that caused her to cast her baby aside. While they were thinking about it—gorillas recycle quickly—Dolly became pregnant for the second time. She was ten years old, and just coming into maturity. Her keepers hoped to give Dolly some basic instructions before the second baby arrived.

Steve Joines, who had been trying to get started with the zoo, was asked if he cared to take on the assignment of acquainting Dolly with the technique of motherhood. A graduate student in physical anthropology and primate behavior, twenty-four at the time, Steve took the assignment eagerly. He made some false starts—for example, showing Dolly movies of gorilla mothers in the wild caring for their infants. When this approach proved ineffective, he recalled that a gorilla will do what you want it to do if it likes you. He had a headstart on this, for Dolly had begun to look upon him with particular fondness. Steve set up house with a "baby" constructed from beige denim filled with foam and painted with an ape's face. To relate that bit of fluff to the next gorilla baby would require a leap in logic on Dolly's part; that could have loomed as a problem, but it didn't.

Steve's objective was to get Dolly used to cradling and protecting this imitation baby, since gorillas do not seem to have a natural maternal instinct. It had of course been assumed that they did have this instinct, but zookeepers were finding out differently. In the wild, each young female has the models she needs in older females who are bearing and rearing children within a family system. It is now speculated that in captivity wild gorillas are at a loss without such models. At the zoo, Dolly found Steve a respectable substitute.

She let him rehearse her in handling the doll, and when her second baby came, she was ready. She used the "mothering techniques" Steve had taught her.

"I understand about the first time," Steve told me six years later in 1981, when he had become one of Belle Benchley's successors in solving puzzles about gorillas. "Imagine what it would be like for you if you were a ten-year-old female gorilla, you had never heard of gorilla babies in your life, and suddenly one pops out. You have absolutely no idea what it is. You're frightened to death."

The significance of this report, from the standpoint of an investigation into interspecies communication—how far it has gone, or how far it *can* go—depends on the fact that it required words as well as demonstration to teach Dolly. Only when he combined relaxed verbal instructions with his demonstrations did Steve begin to make headway. The two-dimensional gorillas on the movie screen, doing a perfectly good job of mothering but with no hints, pokes, or private messages for Dolly, held no interest for her. "She couldn't relate," Steve says.

What she could relate to was Steve's manipulation of the

projector, which she found fascinating. Accepting the clue, Steve provided Dolly with both something to handle and words to clue her in on what it was about. He wasn't concerned that the words might be new to her. Like Belle, he saw the gorilla as a quick study if it chose to be—able to connect spoken sentences, if repeated a bit, with a given course of behavior.

"I taught Dolly to respond with the doll on voice commands," Steve recounted. "She understood the terms and responded well. I would say, 'Show me the baby.' She'd show it to me. 'Be nice to the baby, Dolly.' She'd cradle it on her breast. 'Turn the baby around.' She just came to understand it, the idea and the words, too. When the real baby came, the attitude was right and Dolly was ready. I had to use a voice command only once. That's when her real baby cried. Dolls don't cry—not denim ones. The baby was called Binti, which means something like daughter in Swahili, and Binti let out a squall. Dolly was disturbed. She looked at me as though to say, 'Okay, what do I do now?' I said, 'Be nice to the baby!' And she was nice. A mother on the second try."

Six years of mingling with gorillas and other great apes changed Steve Joines into an advocate of the position that the gorilla is not only "the best critter alive" and "a noble creature" but is better at absorbing human language than humans are at grasping the meaning of gorilla talk. He accepts the Belle Benchley hypothesis: gorillas have an intricate system of communication, and they have it with or without human intervention.

All who work with gorillas are conscious of the possibility of a mishap, not because of a gorilla's maliciousness but because of its strength. Steve was involved in a small incident indicating, to him, that a gorilla will attempt to express a complicated thought in a form a human can understand.

After a twenty-minute workout on one of their joint projects, Steve and Dolly sat down together for a rest. Affectionately, she took his hand. A security guard whom Dolly particularly disliked and distrusted suddenly appeared. Frightened, Dolly squeezed Steve's hand with full gorilla force.

"Dolly, let go!" Steve commanded.

She immediately leapt away to her bedroom. Steve asked the security guard to leave, and Dolly came out, full of amends. She was looking for a way, Steve contends, to let him know that she knew what she had done and that she regretted it. She came up, took Steve's hand gently in her own, and kissed it.

"It's the first time she understood," he said, "that she's a lot more powerful than I am."

Steve, who is as familiar with formal experimental methods as he is with the more relaxed atmosphere of the zoo environment, asserts that a primary source of confusion in current interspecies experiments derives from the fact that the doubting researchers ignore the practical experience of the zoo attendants, who "can be less educated than the primatologist but much more sensitive to the animal."

"Gorillas talk," says Steve Joines briskly. "Gorillas understand English."

Steve had a chance to become one of the trainers of the Patterson gorillas, Michael and Koko. He turned it down, he says, for a reason that will surprise those who resist the notion that serious progress has been made in understanding gorilla talk. Steve rejected a role in the communication experiments because they were already well along, and he felt he had better work on gorilla conservation instead so that there will be some gorillas around, in the future, to communicate with.

TRIUMPH, AND TROUBLE, IN THE WILD

So much has been said here about communications advances made with gorillas housed in zoos and chimps (and other animals) housed in research labs that it is possible to lose sight of experiments—not limited to the field of communication but certainly involving it—that can hold much higher implications for the future of humans and apes in a changing world than any of the more clinical tests are likely to.

Successful efforts by Jane Goodall and Dian Fossey to make camp near wild chimpanzees and wild gorillas have permanently altered our sense of what can and can't be done in terms of encouraging animals to let benevolent strangers view their life and culture. To work as close to the apes as these two researchers-in-the-wild wanted to demanded acceptance by creatures who had never before accepted humans under such circumstances.

While either might resist applying such an easy catch phrase as interspecies communication to their work, it could easily be argued that Goodall and Fossey have achieved the greatest interspecies feats of all. It seems unfortunate that an increasingly narrow emphasis, in laboratories, on word exchange should lead anyone to imagine that unless there has been a transfer of hundreds of words in English equivalence the interspecies quest is not well along.

Jane Goodall arrived in Africa first—in the early 1960s. She worked in Tanzania with wild chimpanzees. Dian Fossey, a Stanford physiotherapist, visited her before making her own attempt with gorillas, starting in 1962. Locating at Mount Visoke, in the volcano regions of the African highlands, Dian studied and virtually intermingled with the mountain gorillas she called friends, gorillas who were not mere shadows of themselves as they sometimes are in zoos. The vocal emoting of gorillas fascinated Fossey, and she imitated everything, including their deepest belches.

When a gorilla in the wild finally touched Dian Fossey's hand—what mighty revolutions of the heart must have preceded the decision to do this!—it was a revelation for many persons. Even though they might envision a day when gorillas will be eager schoolchildren, those who watched from afar could experience wonderment that the human-creature and the ape-creature had been able together to cross a boundary so great.

If an ape has not experienced humans before, its first contact with a human must be something like a human's encounter with beings from outer space. As often as science-fiction writers have cast such scenarios, we really don't know if humans would react to a visit from the unknown with greater or lesser reluctance and fright than the apes have shown to their human visitors.

Neither Jane Goodall nor Dian Fossey was attempting to deliver language lessons to the higher apes. Both depended for their survival on mastering a sufficient number of signs, gestures, and vocalizations to get through difficult, unpredictable circumstances. Dian Fossey can't listen to gorillas vocalizing with each other and make a report on the latest gorilla gossip like someone reporting on the chatter at a party. She did find out what makes an approach to gorillas far easier: imitating the sounds they make and duplicating as precisely as possible their own vocalizations. (Ideas spread. Only a few years later, an ocean diver named Debbie Ferrari was able to apply the same idea to whales; she imitated their own sounds to bring them in closer.)

After two thousand hours of watching gorillas, Fossey came forth with the not too surprising report (it might, however, surprise eaters of gorilla meat) that she had seen "less than five minutes of what might be called aggressive behavior."

Yet the gorillas—Dian Fossey's jungle friends—have been the object of vicious human aggressive behavior in a tradition that has continued from Paul Du Chaillu's time to the present day. The great adventures in understanding and interspecies friendship undertaken by Goodall and Fossey have turned into nasty predicaments. In each case this happened not because ape and researcher had a falling-out but because warrior-humans disregarded the tender excitements of learning to see how the world is through apes' eyes. The Goodall experiments were disrupted by kidnapings, the Fossey trials in interspecies friendship by murderous poachers.

By early 1979, an ape craze of smashing destructiveness had all but pulled Dian off the experiments that brought her to international attention. Habitat for the gorillas was being destroyed by the invasion, and Dian spent most of her time fighting the nightmare of increased tourism and increased poaching. In a much-noted incident, she had once frightened gorilla poachers off by donning a Halloween mask, but it hadn't kept them from coming back. Over and over through the year, she and her assistants dismantled snares intended for bushbuck and antelope because they catch gorillas, too. The poachers wrought a notable tragedy that shocked the followers of her famous Digit, a gorilla whom readers in America and Europe felt entirely familiar with from the many popular reports about her work with him.

Jim Doherty, executive editor of *International Wildlife,* reported on Digit's last stand: "Less than three miles from Fossey's primitive base camp on the Rwanda side of Visoke, six poachers stumbled onto a group of 14 gorillas. Startled, the female apes took flight with their young, escorted by all but one of the five adult males. That one, a burly 13-year-old, stayed behind while the other animals made good their escape. Evidently [Digit] the lone defender put up a terrific fight. He received five deep spear thrusts but still managed to kill one of the poacher's dogs before he died. After hacking off the gorilla's head and hands, the poachers fled, leaving the corpse in a small open area where it was found by one of Fossey's assistants."

While it is possible to see Digit's massacre as a random act by poachers who took whatever they stumbled on, Kenholm Stott guessed that there was more to it than that. He suspected the poachers had known about Fossey's special attachment to Digit and knew what act of theirs would cause the greatest shock. Stott believed the poachers had stalked Digit to behead him as an act of deliberate terrorism.

"Is Virunga going to be a national park or a poacher's playground?" asked Dian Fossey.

She antagonized officials with her own antipoaching patrols and complained bitterly that the promoters of tourism were plotting to drive her from Rwanda. Richard Reinauer, an associate producer of TV's *Wild Kingdom,* told Doherty, "It's all so very sad. Dian is striving almost singlehandedly to save the last vestige of an animal species that has been on this planet for thousands of years. In spite of our knowledge, our humanity, our advanced civilization and communications, she is fighting a losing battle and there is little we can do to help."

At Stanford, where the sight of two gorillas working the Coke machines and getting lessons on a computer from Penny Patterson and her colleagues were an odd sort of counterpoint to the ferocious action in the jungle, another kind of battle was in progress. Things were by no means as bad as in Rwanda, but the gorillas were unable to hold their own against a bureaucratic attack, which seemed mysterious, considering their celebrity and the enormous interest in finding out whether the gorilla communicators were really onto something or not.

The university moved in a relentless fashion to remove the gorillas and their trainers from the Stanford campus. The mystery of the university's maneuvers was readily solved, as it turned out, but it had an odd and unexpected connection to happenings in the African jungle.

9

Fireworks Child, the Uncorrupted Gorilla

THE SIGNER, WANTING TO PAY THE GORILLA A COMPLIment, signs "pretty brown eyes." Three words conveyed in sign language.

The gorilla answers, in the same language, "Fake."

The signer repeats her compliment.

The gorilla answers, "False."

It has happened more than once with nearly the same result each time. If the human signer continues an effort to persuade the gorilla to accept the compliment and admire herself, the gorilla answers, "Not true" or "Not real."

This sounds like it could be an episode from Michael Crichton's thriller, *The Congo*, about a sign-language-speaking gorilla named Amy who goes on an African trek to find the lost city of Zinj. Not so. It was Koko, a 140-pound gorilla who was seven years old at the time the gorilla language experiments were hitting their stride in 1978. A companion gorilla, Michael, was a year younger and weighed less, although he had the prospect of growing into a 400- to 600-pounder. Even at such a tender age, the gorillas had five times the strength of humans in their own age bracket or older—and their respective vocabularies outweighed that of any nonhuman who had ever lived.

Like George M. Cohan, Koko was born on the Fourth of July. Her formal name is Hanabi-Ko, which means Fireworks Child. Koko was a fact of American life before Crichton's thriller appeared; in this case, art has followed life. Koko has not gone looking for lost cities, but she would probably be willing to do so if it were suggested. She lacks assurance in respect to whether she is pretty. In nearly all other ways, she is a perfectly self-confident gorilla.

"Poor Koko, she does have a problem with her ego, mainly on the subject of her attractiveness," said her chief trainer, Penny Patterson. "If *I* tell her she's pretty, that's okay. But if someone else tells her, she disagrees."

Like Koko, Penny has a more formal name—Dr. Francine Patterson, Stanford doctoral program graduate in developmental psychology. I never doubted that if she could formally establish only a part of what she so casually cited to me, she was sitting on top of the most significant animal experiment ever conceived.

Penny Patterson is more serious than her casual ways implied, and she is not at all casual when it comes to citing chapter and verse on the experimental principles underlying her research. She has a gift, though, that the behaviorist often lacks—she knows how to surround gorilla studies with an air of informality and spontaneous fun. Those who have had an expert teacher in kindergarten will know the kind of skill involved—an enormous talent for making the difficult seem just the thing to try.

At Stanford and later at the Gorilla Foundation, this was the atmosphere Dr. Patterson was creating. (The formality of the title always rings oddly once you've gotten to know her.) By nature or design, Penny surrounded what was in reality an epic exploration with all the glow of setting off for Treasure Island. For gorillas who are only dimly aware, if at all, of scientists scribbling down their every utterance, the sense of doing something new and interesting must never let down. The training had to seem almost as natural as if they were out there in a gorilla family in the wild, traipsing toward fresh adventures. That everyday, look-how-natural-we're-all-being atmosphere allowed Penny and a number of deft assistants to push the experiment into new territory where no language experiment had ever been before.

The path has been anything but easy. Penny has received many bites, delivered by gorillas still in their babyhood who communicated their annoyance elementally. She has survived these assaults and even more ferocious bureaucratic assaults. She has organized one operation after another to save her grand idea of working with gorillas. She kept the gorillas from being wrenched away from her after the experiment took flower, and she has survived academic gamesmanship and intramural jealousies. She has raised money for the gorillas' upkeep, and she has searched for a larger home for Michael before he becomes a 500-pounder and bursts through the walls of the trailer-house he has lived in since Penny scrambled to import him from Europe.

The bruises from this struggle are not inconsiderable. Still, having survived the early rounds of gorilla bites, Penny can probably survive anything.

Penny first began trying to beg a gorilla from the San Francisco Zoo in 1971, but it took a while to accomplish her purpose. Karl Pribram, the distinguished Stanford brain scientist and prober into animal mentality who had that "mishap" with Washoe, helped her in her early work and was with her on her first visit to the zoo.

Penny had always had a fairly clear idea of where she was headed. "My patron saint is St. Francis, who spoke with animals. As a child I thought, 'Oh how neat to do that. If I were a saint, that's the talent I'd want.' "

When the Gardners, the couple who wanted to create an "overlap" between humans and animals, came to the Stanford campus in 1971 and described their experiments with Washoe, Penny suddenly realized how she could fulfill her childhood ambition: she would take up Ameslan, study how to convert it into sign language for animals, and follow the Gardners. She would address "the ultimate question with the ultimate animal." At the time, she was thinking that the ultimate animal was the one the Gardners had picked, the chimpanzee; but then she saw a baby gorilla at San Francisco Zoo.

All baby gorillas are a piquant mixture of mystery and moodiness. Penny has never doubted that she found "sweet lovableness" in the infant Koko. It was evident even when Koko was just a tiny sprig of a gorilla clutched to a large female on the day that Pribram and Patterson came begging at the zoo.

You should understand that a gorilla is no small beg. At current (1981) prices, a gorilla with class could command as much as

$30,000. A pair recently went for $50,000. Koko and Michael, who have mastered skills no other gorillas have ever attained, are now priceless, and anyone who visits them must be x-rayed first to make sure he or she is free of tuberculosis, to which gorillas are peculiarly vulnerable. Mrs. Benchley's Ngagi, a seemingly powerful five-foot-eight-inch 501-pound specimen, died when he was less than eighteen years old. Mbongo, who topped 600 pounds, died at fifteen years and five months. In the wild, gorillas are believed to live an average of twenty-five to forty years, but some have lived to sixty. Keeping gorillas alive in captivity is no easy matter, as their chances of being felled by stray diseases are considerable.

When she first tried to get a gorilla, Penny was unsuccessful, but she didn't give up. The San Francisco Zoo wasn't selling, giving, or lending, and they didn't want Penny in the gorilla grotto conveying sign language to any gorilla who would pay attention. Even so, Penny kept thinking about the baby gorilla, determined that it would be her subject. When she came back nine months later to work on the zoo people again, she had an assist from that susceptibility of gorillas to illness. Dysentery was running through the gorilla compound, and Koko's mother was having trouble producing milk. As a health measure, Koko was sent away with Penny—not for keeps, just on loan.

Koko's first significant act was to bite Penny on the leg, and she kept on biting. At first, it was all bites and attempted bites. Installing Koko in a trailer-house on the Stanford campus, Penny kept her nerve and decided to outlast a gorilla who was small but had lots of fight. In order to teach Ameslan to Koko, Penny firmly shaped the ape's reluctant fist into the sign to be learned. The idea was to repeat, repeat, and repeat, relaxing the grip as time went on until the animal was making the sign on its own. If you don't mind how many gorilla bites you take home at night, this has now been proved a workable practice.

The standards for including a word in a gorilla's working vocabulary are necessarily strict. A word is only considered learned (behaviorists call the process *qualifying a sign*) if the gorilla uses it fifteen days out of thirty or fourteen days in a row. (Weaving such phrases as "savoir faire," "ennui," or "gentle herbivorous anthropoid" into occasional conversation just doesn't count.) A veritable mountain of data is prepared to document each time a word is qualified.

When the sign-language lessons had gone far enough, the trainers could ask Koko what she was afraid of and get an answer like "Hats . . . dogs." What did she think was funny? "Clowns . . . bugs." What did she think was pretty? She pointed to the hair on her stomach. What could make a gorilla happy? "Gorilla love Coke." That answer checked out with experience—Koko loved Coke and loved to do her own manipulations with Coke machines.

When somebody asked her what was boring to her, Koko's answer was "body parts." Penny interprets this answer to mean that trivial questions are boring to a gorilla. "She keeps getting asked about her head, her eyes, her bellybutton. She knows all that so well that by now she's bored with the subject. I don't know what the limitations would be on what she could learn. It seems to me she learns as much as a child, just a little bit slower."

Penny has no snobberies about the appropriate curriculum for gorillas. Instead, she let the gorillas develop their own feeling for what might be of special interest. And Koko, as it happened, was drawn to the image of Count Dracula.

"She *especially* likes Count Dracula," Penny commented in the rambling style that suits the mood of the Gorilla Foundation. "He appears in *Sesame Street* magazine, which Koko enjoys, along with *National Geographic World*. We give her magazines and books every day. She'll spend ten or fifteen minutes with a magazine, going over the pictures. She calls Dracula *bird*. Sometimes she calls him *frowning bird*, referring to his scowl and that bat that flies around."

Some word signs are easier for some gorillas and harder for others. "A really hard word for Koko to sign is *hit*. Michael likes to sign *hit*. Koko knows it but she will rarely sign it. Michael likes to sign *squash*. He likes to sign *strangle*. He likes to sign *bite*. He likes words that are aggressive. I don't know yet if that's because he's a boy. He once saw a picture of a coyote attacking a lamb. After that he talked about *biting—dogbiting—being red*. Another time he began talking about something he *hadn't* seen a picture of: *cat eating birds*. He would bring it up with five or six different people in a day. *Cat eat bird* was very much on his mind. We asked him what he was talking about, but we still don't know for certain what brought on this image. At Stanford, cats were certainly all over the place, probably lab escapees. Baby cats were being reared around the trailer, and the cats must have to hunt for their food. So maybe Michael saw something. If it's violent, he remembers.

"Some of Michael's expressions are clearly made up. Like when he says *Alligator bite lip.* Now *alligator* is one of his favorite ideas, which he's gotten from pictures, but *lip* as it's used here is a gorilla invention. The gorillas use the word *lip* to mean girl or woman and *foot* to mean man. Koko started using the word *foot* to refer to a man named Al who worked across from us when we were at Stanford. Al used to tickle her feet, so she started calling him *foot.* Then all attractive men became *foot. Lip* came about in a somewhat similar way. *Lip* originally referred to women visitors wearing lipstick. Koko always immediately knew if they wore lipstick, and she would 'comment' about it. I don't wear a lot of lipstick. But then she started calling all the women *lip,* including me. She knows I'm Penny, but if she's feeling lazy, she just says, 'Lip do this. Lip do that.' We certainly didn't teach Michael either *lip* or *foot* for woman and man, but he uses them, and they must have come from Koko. *Alligator bite lip* means 'The alligator bit a woman.' "

The gorillas' creative linguistic constructions also turn in the direction of epithets, as the need requires, of course. In the October, 1978, issue of *National Geographic,* which alerted the world to a claim that the talkingest gorilla of the age was coming into her prime, there is a series of photos which illustrates that gorillas share chimpanzee scatology. Penny had set out to reprimand Koko for joining Michael in a game of bite-the-legs-off-the-Raggedy-Ann-doll. Michael got one leg but Koko got the other. She was feeling at least half innocent of the crime, but isn't keen on being reprimanded, so she said to Penny (and the photos follow it, sign by sign): "You—dirty—bad—toilet."

Penny agrees with that finding from the chimpanzee experiments: "Gorillas like bathroom words. That's something they hang onto. They have a large lexicon of words like *stink, dirty, rotten, bad, obnoxious.* It's the sort of word they can remember. Koko doesn't insult me quite so much anymore. She's mellowing. The person she does it to now is her photographer, Ron. She calls him a *devil toilet*, perhaps because she resents having to be on good behavior for the camera."

Hurling insults, inventing colloquialisms, blandly discussing her emotions—these are just a few of the achievements that have brought Koko such distinction. Over the course of the last decade, Koko could have made the *Guinness Book of World Records* almost every day for one accomplishment or another. By the trainers' account, these were not impossible accomplishments for gorillas, just, from the human standpoint, firsts. First gorilla to master fifty human words, then 150, and then 300. First gorilla to accept questions from the press. First gorilla to receive a Mao cap as a gift from admiring Chinese. First gorilla to try skateboarding. (Michael may have been the first gorilla to ride in a supermarket basket.) First gorilla to cruise the California hills in a red Datsun, although Koko didn't drive.

Most of these feats, including the vocabulary signing, are there in photographs or on film to be examined. Fireworks Child has had her every passing remark more completely documented than Samuel Johnson at his wittiest.

INTELLIGENCE TESTS AND SOMETIMES CHAMPAGNE

One of the ways to demonstrate Koko's objective knowledge is through the administration of IQ tests.

Is Koko, by gorilla standards, a genius? Did the Patterson crew hit it lucky with her and the younger gorilla, Michael?

We'll have to talk with more gorillas before answering. The usual human-oriented Stanford-Binet scale IQ tests have recorded Koko as high as 95 and as low as 70. An average human IQ is around 100. This would indicate that, compared to a human, Koko has tested out both slightly below and well below the human average. According to Penny, Koko would do a little better if there were not a "cultural bias" in some of the test questions—for example, a question on foods that are good to eat pictures a block, an apple, a shoe, a flower, and an ice cream sundae. Koko picked the flower.

This lowered her IQ rating because the person administering the test is required to rate that a wrong answer, but for gorillas it is right—they eat flowers. This hasn't led to any great movement for change in IQ tests or their scoring ("Let's Be Fair About Gorilla IQs") but it shows how hard it can be to arrive at an estimate of gorilla intelligence.

How does Michael rate on the gorilla-genius scale? Penny guesses her apes fall somewhere within the range of apes-as-usual. The main difference is that not all apes have the opportunity to pursue higher education in American academe. (I write this knowing that it involves a cultural bias and that the real higher education, for gorillas, probably remains in the jungle itself.)

"My guess is," says Penny, "that we don't have dummies from the gorilla standpoint and we don't have geniuses. Neither Michael nor Koko is a genius—the odds are too much against it." And she mentions what Belle Benchley had found: "Gorilla learning, more than you'd probably think, has to do with motivation. The quickness, or lack of it, with which they learn is not so much intellectual limitation as a motivational limitation. When they're interested, they learn immediately, and they remember. One trial is all it takes.

"On New Year's Day last year, someone brought Koko just the tiniest bit of champagne in a bottle. And that's not the custom here. Koko was given part of a martini when she was just a little thing, and I learned right then that a tipsy gorilla is not what you want. The martini made her hyperactive and obnoxious. But there was some acting-up on New Year's, and she had just this tiny touch of champagne, and never forgot it. We had told her 'This is champagne,' and she signed *champagne* to get some. Now she talks to us about champagne: 'Want champagne.' Champagne's a big subject with Koko. She knows what she likes."

Although Koko has been tested to some degree on at least 600 different words, her actual working vocabulary—words she uses with some regularity—is, according to Penny, in excess of 300 words, which is large. In 1980, Michael had a vocabulary of 250 words. If Penny is correct in thinking that part of Michael's vocabulary was taught to him by Koko as well as by the humans around him, this was at least as significant as the vocabulary Koko learned.

Transfer of language (ape-to-ape rather than human-to-ape) had been a primary goal of the interspecies communicators.

The next big question is whether Koko and Michael together will transmit sign language to their children. But before it can happen, they will have to have offspring. "Koko is really trying but Michael just isn't there yet," says Penny. "We're guessing that he's about two years too young to know what's going on. She pushes her bottom close to him, and she alway presents her back to him when they're playing. Or she'll go down on her elbows with her bottom sticking up in the air. Michael ignores this, although he sometimes makes pelvic thrusts on a black tub, which would be the closest thing in shape to a gorilla. But he seems to make no correlation between his tub idea and Koko's idea."

While Michael is staying dumb about the needs of female gorillas, Koko has apparently begun to take family matters into her own hands. She has largely given up her personal choice of swear words in favor of talking about dolls—sometimes to herself, sometimes to Penny, sometimes to the dolls. "When she volunteers conversation, it's usually about dolls now," says Penny. "She has all kinds—human babydolls, gorilla babydolls, a basic King Kong doll—that she plays with. If they have movable hands and arms, she manipulates them. She makes the dolls sign for her and she signs to the dolls. And she has signed *soft* to the doll. She kind of resents me watching her at this, so I have to pretend I'm busy with something else."

It will never be recorded with the conclusions that reach the computer, but it could almost be gathered from this that Koko has conceived an ambition. Perhaps she wants to follow in the footsteps of her role model and become a language instructor, another Penny Patterson.

A QUESTION FROM THE JUNGLE

What is the best way for a human to approach an ape in the wild? Jane Goodall sent this question back from the Gombe Research Centre in Tanzania for Koko to answer. It was not that Jane didn't know the answer; she was accustomed to teaching her students that sitting or crouching is the best way to carry out research near the great apes. But, as she wrote Penny, "I'd love to be able to tell the Gombe chaps straight from the horse's mouth, as it were."

When Penny asked Koko whether people should stand up or sit down when watching a gorilla, Koko's reply, she claims, was a strenuous: "Down!"

Like all the questions, it was repeated to make sure the gorilla had understood. "We asked her three times, and she said *down* each time," reported Penny. "The third time she went right down on the floor. It was like she was saying to us, 'Don't you get it?' "

How does Koko make distinctions between the notions of should and shouldn't? As far as can be judged, the gorilla's morality—its sense of should—is absorbed through the pores from the actions of the people around it, in much the same way it is absorbed by a human child. Praise and remonstrance play a role in this, but why a gorilla should have some innate code of the jungle dictating that jungle callers *should* throw themselves down before the apes is more mysterious, and Koko's answer can't be readily explained. It is not taught in the trailer-house that newcomers must bow down either to gorillas or to Penny when they enter.

No matter how such seemingly innate knowledge originates, a gorilla knows what it knows and doesn't like to be contradicted. Once the gorillas had learned to sign, they were stubborn about admitting that an answer could be in error. Years ago, Penny described how Koko won an argument with her. She had shown Koko a piece of white cloth and asked her to name the color. "Red," said Koko. When Penny tried to correct her, Koko wrangled. Finally, the gorilla pointed to a tiny piece of red lint to prove she was right.

Michael has his own notions about the meaning of the color red. "Red mad gorilla" is a frequent phrase of Michael's. The trainers believe he chose the combination of words himself and interpreted the phrase to mean *angry;* they believe he uses *red* to underline the intensity of his feeling. He could also apply the phrase to others. When he felt someone was acting in an accusatory or angry manner, he would sign "Red mad."

"People do turn red when they're angry," Penny said. "I think Michael says *red mad gorilla* because he has seen people turn red, and knew it meant anger: maybe he thinks from that that he turns red when he's angry, too. It took us time to understand that, in Michael's vocabulary, red can mean anger. Now we know it and we see him using it where it fits."

Koko's contrariness in serious disputes about color and other subjects has its own charms. In the Penny-Koko household, this tendency also became a weak spot by which she could be manipulated. Ron Cohn, the "devil toilet" photographer, tried to make Koko stop bending and breaking spoons, which only inspired her to step up the pace of destruction. "Good, break them," signed Ron—which caused Koko to start kissing the spoons.

Playful gorilla destructiveness can be a problem. A few experimental visits to Penny's nearby home resulted in a broken bed and other bashed items and the trips had to be abruptly stopped. There were just too many things there for Koko to examine and wreck. But the trailer-house was a large world, adapted only in minor ways to gorilla control and gorilla preferences. Chain-link barriers protected windows, a cost-saving precaution against gorilla enthusiasms and upsets. The regular gorilla chore in the jungle of making a nightly nest from branches and twigs was unnecessary at the trailer-house, as was the practice some gorillas have of establishing that nest ten or twenty feet off the ground. But Koko wound up with a permanent bed she liked made from a motorcycle tire draped with rugs.

This led to one of those small incompatibilities in gorilla life that can occur in the suburbs but not in the jungle. Michael eats rugs. So now at night he and Koko are together but apart in adjoining sleeping spaces separated by a chain-link barrier. They can still fondle and touch, which is desirable because one day Michael will figure out why Koko becomes urgent and presenting when she's in estrous. They can make signs and pass on what they know, but Michael can no longer eat Koko's comfy bed.

What information do gorillas pass back and forth? Anecdotes passed on by the gorilla crew suggest that the gorillas are gossips on events of the day.

Gorillas sign, answer questions, dutifully take tests, repeat (sometimes with irritation), and feel and express anger—and they can recycle that anger into angry word-venting—but they also report and volunteer. What they report on can be actual happenings or possibly they can be dreams, a visualization of what *might* happen or what they would like to happen.

Gossipy gorillas are just what the world might order if it could have anything it wanted. Gorillas who can tell you the color of a spoon or what they like to eat aren't telling you much that you couldn't find out otherwise. But gorillas who see and tell may have matters of considerable moment to report when their mastery of that several-hundred-word vocabulary becomes flexible enough.

In the case of Koko and Michael, the ability to gossip reportedly was growing. One day Barbara Weller, a volunteer with the Gorilla Foundation who became deeply involved with the language experiments, came to find out what Penny or Ron might know about "a red-haired girl" and "big trouble." Penny related the following: "We were in the lab at Stanford when a girlfriend of one of the workers came in, really storming. She started a violent argument—screaming, slamming doors, it was awful. Barbara was the only one who hadn't seen what happened. I saw it and the gorillas saw it. Koko and Michael watched, very concerned about the whole situation." Afterward, the gorillas reported the "big trouble" to Barbara, complete with an understandable reference to a red-haired hellion.

CREATIVE LYING

Had Penny Patterson and her instrumental co-workers merely been the developmental psychologists who claimed to have given the world its first talking gorillas they would have caused a great stir in the annals of animal experimentation. But one early observation by Penny changed the dimension of what she was doing and upped the stakes.

"The gorilla lies," Penny said.

That was a direct confrontation of the assertion that even though animals may signal ingeniously, they are not capable of creative thinking. Whatever the objections to them, lies are creative. Indisputably, they fall within that part of language which is "half art." Lies are different from memorization or mimicking. They call for a conclusion on the animal's part that it would not like to get caught or that it would like to talk you out of something.

Behaviorists who were having trouble accepting the Patterson conclusions considered the allegations of gorilla lying among the most suspect of the various claims. Molecular evolutionist and nuclear-medicine specialist Dr. Jerold Loewenstein, one of the scientific sleuths who was trying to make progress in solving evolutionary puzzles, was not a supporter of Penny Patterson's experiment in other ways, but he was not as skeptical about an animal's "creative lying" as many scientists are. He accepted the assertion matter-of-factly.

"Dogs lie," Dr. Loewenstein said. "Cats lie. You come down the stairs, a dog leads you into feeding him, and he doesn't give any sign that your wife already fed him—he just told you a lie."

As Penny described them, Koko's lies took a more direct form. She would accuse the innocent if that seemed a handy escape. These accusations were reportedly made in Ameslan.

Penny described an incident in which Koko "was caught in the act of trying to break a window screen with a chopstick she had stolen from the silverware drawer." Asked what she was doing, Koko replied, "Smoke mouth." She put the chopstick in her mouth to do some of her pretend-smoking, a game she likes to play. The incident suggests that, aside from lying, gorillas also steal from the silverware drawer and conceive fancy ideas for jimmying windows.

The kitchen sink gave way a few inches one day when Koko sat on it. This was scarcely her fault, but she seemed to feel guilty anyway. When Koko was asked, "Did you do that?" her reply, according to Penny, was to fix the blame on Penny's deaf assistant, Kate. "Kate there bad," signed Koko.

Koko's trainers say that along with the lies come confessions. The day after Koko returned to an old habit and bit Penny's hand, when Koko was asked what she had done the day before she reportedly signed "Wrong, wrong." Three days later, seemingly still ruminating about this, she signed "Sorry bite scratch." The bite on Penny's hand had changed appearance as it healed, and resembled a scratch.

"Wrong bite," signed Koko.

"Why bite?" asked Penny.

"Because mad," Koko signed.

Koko had come a long way from that infant stage when she was an inveterate puncturer of Penny's skin—with no apologies.

CORRUPTED AND UNCORRUPTED: WHAT IT MEANS

Is a gorilla who can lie, insult, deceive, and gossip still a gorilla in its basic, uncorrupted sense? Many would say no, that Koko has been debased by her exposure to human behavior and values. What does it really mean for an animal to be corrupted?

An animal sits in a cage. It was born to roam the forest, to make a nest at night, to live in a family of older and younger members, to expend effort in acquiring food, to have daily encounters with other animals—and even to wonder about them, to go through an hour, a week, a month, or a year in a spiral of changed times and new adventures. But the animal sits in a cage. The times do not change for this caged animal, although its hair may change to white like yours or mine. A keeper coming with food, wary of the captive, knowing that a finger unwatched can be a finger missing—that is the caged animal's big adventure of the day. Or the distant, loony faces of people peering and going through conniptions to make the animal focus its attention on them. Seldom does this have in it the spirit of adventure. The animal sits in its cage. It is totally corrupted. Life has made it different from what it was meant to be.

Those experimenters who isolate the "test animals" often believe that isolation protects the experiment from bias and that the use of multiple contacts for one animal prevents the buildup of human bias in a test administrator who simply *wants* the animal to succeed.

But isolation is also corrupting in that the animal becomes someone other than itself. This can be demonstrated with a thousand examples or in any zoo.

Failure to encourage the animal's growth through affection or familiar contact with a dependable face is corrupting.

Nearly all the successful language experimenters have been careful to give their captive animals a larger life than a captive would normally have. They know that word exchanges seem to work best with animals who have, if not their original wild environment, some version of a full life.

The life that Fireworks Child led when she was a fully accepted tribal member of a very busy university crew with much to do was as different from life in the forests of Mt. Visoke as it could possibly be. But her life had something a cage life doesn't have: wonderful variety, mixtures of attitude, changing scenes. She had a nickname, car rides, an occasional chance to swing in the trees on the Stanford campus. She had press conferences, a flow of visitors who were different than ordinary gawkers, a mixed diet of common and exotic fruits and vegetables. She had a corner to stand in, to show penitence, when she had been bad. She had a ranch up in the hills—the 1,300-acre Djerassi spread where gorillas were welcome—to exult in on holidays from the old school grind. She learned to put money in the campus drink machines, and she could be depended on to point the way when a drink machine was in sight. In those great moments when she had 1,300 acres—or the rest of California, for that matter—to escape to if she chose, she may have felt as free as if she had never left Africa. She was still a gorilla, and she was still a captive gorilla, but the terms of her captivity were different than they were for others who had come under the hands of humans. And even if she worried about her looks and didn't believe the flatterers, she had a mighty sense of self.

She had escaped corruption through the breadth of the life she led, the great sociability of the life that Penny made for her. As it happened, she lied, insisted, had stubborn fits, corrected her correctors, and hopped on the Dracula bandwagon when there were so many better fellows in the world, but these were all signs of a free will, not corruption.

Rigidity is corrupting.

A bit of champagne on New Year's is life-enlarging, even for gorillas.

10

The Gorilla Dialogues

ALTHOUGH PENNY PATTERSON WAS THE FIRST INTERrogator of gorillas, Barbara Hiller soon joined the experiment and became the second of the two "surrogate mothers" to gorillas. (She is sometimes referred to as "Michael's grandmother.") Her style of questioning the apes is as engaging as Penny's, often taking the animals on odd trips into the wonderland of words.

The "gorilla dialogues" reproduced here are from the Gorilla Foundation. They are usually abridged or condensed, because the gorillas' verbal ramblings don't always lead somewhere.

The opening conversation explores how a gorilla approaches life emotionally. I am not presenting this as a historic conversation—their trainers believe that gorillas have had better things to say than this. But it is an example of the extent, and the curious style, of a supposed conversation between a human and a gorilla.

The interrogation was conducted in sign language, and the dialogue has been extracted from questions put over a several-day period. As the surrogate mothers point out, gorillas need time out for thinking about coyote attacks, dealing with visiting Ph.D.s, and scanning the new magazines (they look at the pictures).

BARBARA: What do gorillas say when they're happy?
KOKO: *(Grins and beats her chest.)*
BARBARA: What do they say when they're sad?
KOKO: Frown.
BARBARA: What do gorillas say when they're mad?
KOKO: Think blow.

BARBARA: They blow when mad?
KOKO: Mad gorilla.
BARBARA: What do they say when they're happy?
KOKO: Me happy. Monkey.
BARBARA: You talk monkey?
KOKO: Gorilla me monkey.
BARBARA: Okay, what do monkeys say when they're happy?
KOKO: Smile.
BARBARA: What do gorillas say when something is funny?
KOKO: Gorilla funny.
BARBARA: What's scary?
KOKO: Gorilla.
BARBARA: Are you scary?
KOKO: Koko good.
BARBARA: Is Mike scary?
KOKO: Know.
BARBARA: Is he?
KOKO: Mike Koko good.
BARBARA: Do you know any bad gorillas?
KOKO: Gorilla Koko good.
BARBARA: You teach me gorilla?
KOKO: Love Koko hurry.
BARBARA: What do gorillas say when they're happy?
KOKO: You know . . . smile.
BARBARA: What do gorillas say when mad?
KOKO: Koko knows . . . good Koko.
BARBARA: Well what?
KOKO: *(Beats her chest.)*
BARBARA: What do gorillas say when they're scared?
KOKO: Hurry drapes. *(This would appear to be an adept answer. Koko, the experimenters maintain, orders the drapes pulled pronto if something outside the window at the Gorilla Foundation scares her.)*
BARBARA: What scares gorillas?
KOKO: Trouble.

Trouble is a word that, according to Penny, Koko has used often and seems, deep in her gorilla heart, to understand very well. In the early stages of the dialogues, Penny once confronted Koko quite angrily—Penny knows how to scare a gorilla—with a torn sponge. Koko was surely the culprit and knew her own guilt. "What this mean?" Penny demanded. "Trouble," said Koko, sensing exactly what was up.

Gorilla dialogues, as presented by the Gorilla Foundation, are a muddle-wonderful—a bit touching, a bit comic, a bit confused, highly cryptic, highly engaging; they're not quite like any other conversations reported from anywhere. Before his chief trainer started to disown him as a maker of sentences, a conversation with the chimpanzee Nim went like this:

TRAINER: Nim! You bad!
NIM: Nim sorry.
TRAINER: You very bad.
NIM: Me sorry. Hug me.
TRAINER: Come here.
NIM: Hug me. Hug me.

This is very humanlike and quite childlike. It's not at all gorillalike. Who knows what a gorilla means when she says, "Think blow"? How did she choose the words? It is very strange stuff, gorilla talk. We guess what the gorilla means, and lose it, and guess it again, and lose it again; it is something like a shadow dance. The gorilla retains some of its remoteness even in Ameslan but confirms what gorilla keepers everywhere were already deducing—that chest-pounding can mean either joy or anger.

If there is a great deal to be found in the chest-beating interview and other of the gorilla dialogues that is anthropomorphic (the fear of so many behaviorists), it would seem to lie mainly in the gorilla's insistence on how good she is, how good Mike is, and what perfectly right-thinking beings gorillas are.

In Koko, this trait was consistent. When Penny Patterson reported that Koko had seemed to appraise herself with the words "Fine animal gorilla," many of those who have looked for communication with animals took note of those words.

"Fine animal gorilla" has an almost unforgettable ring to it.

"Love Koko hurry" also has a ring to it, and seems quite clear—there is an emergency need for a show of affection; please deliver.

All Gorilla dialogue photos by Dr. Ron Cohen

FIVE CONVERSATIONS WITH SOME TALKATIVE GORILLAS

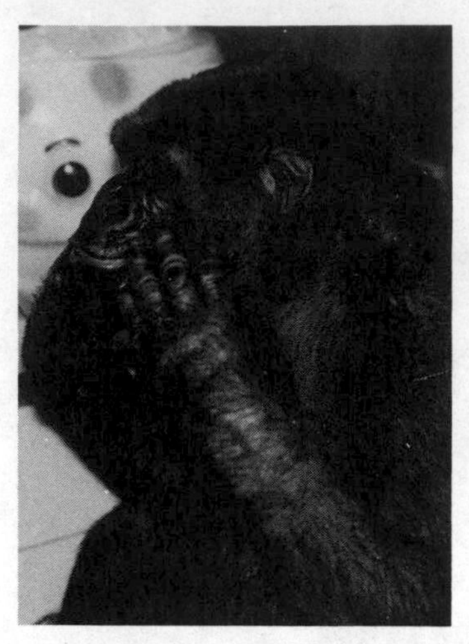

**1. Comedy Patter?
Or Is This Gorilla Confused?**

Penny Patterson and her chief assistant, Barbara Hiller, like to credit their charges with having a sense of humor. Finding out that Barbara had a false tooth, Koko christened her "Faketooth." You can see in the following exchange either a very hip or a very confused gorilla.

BARBARA: Would you like to be able to fly like a bird?
KOKO: Down.
BARBARA: You'd rather stay on the ground?
KOKO: Down floor.
BARBARA: I think you're smart.

* * *

BARBARA: What do people put on their hands when it's cold?
KOKO: Stethoscope.
BARBARA: Koko, that's weird.
KOKO: Think funny.

* * *

PENNY: What is a problem?
KOKO: Work.

* * *

PENNY: What does *wrong* mean?
KOKO: Fake. Koko good that. *(But the good seems to refer to her breakfast.)*
PENNY: What does *wrong* mean?
KOKO: Bad.
PENNY: What does *stupid* mean to you?
KOKO: Koko lazy devil.

2. An Elephant Kind of Thirst

One day Koko used "a fat rubber tube" to drink with. Her trainers were never quite sure if this reminded her of an elephant's trunk, but thereafter Koko seemed to find it convenient to claim she was an elephant when she wanted a drink.

KOKO: Sad elephant.
BARBARA: What do you mean?
KOKO: Elephant.
BARBARA: Are you a sad elephant?
KOKO: Sad . . . elephant me . . . elephant love thirsty.
BARBARA: I thought you were a gorilla.
KOKO: Elephant gorilla thirsty.
BARBARA: Are you a gorilla or an elephant?
KOKO: Elephant me me . . .
BARBARA: You want a drink, good elephant?
KOKO: Drink fruit. *(Later, pointing to a can of soda, then to her glass.)* That there.
BARBARA: Who are you?
KOKO: Koko know elephant devil.
BARBARA: You're a devilish elephant.
KOKO: Good me thirsty.

3. Strangle Talk

Michael and Koko both give serious inspection to animal pictures in the magazines they see on a daily basis. Photos of a coyote, including a picture of a coyote biting a lamb on the neck, made a lasting impression. His trainers said that Michael declared the coyote to be a dog, and that he ruminated over what he had seen. The jumble of language coming out of him included "Cry sad," as though he might be worried about the fate of the lamb.

BARBARA: Do you want to tell me more?
MICHAEL: *(Looks into distance. A few minutes later, Barbara hears the sound of gagging, a sound Michael had been making while talking about the dog-coyote.)*
BARBARA: What are you doing?
MICHAEL: Strangle.
BARBARA: Why strangle yourself? What do you want to strangle?
MICHAEL: Strangle.
BARBARA: Why are you signing strangle? You want strangle someone? Something? Who or what you want to strangle?
MICHAEL: Gorilla more hit.
BARBARA: Hit? Strangle? What are you talking about? Why hit? Why strangle?
MICHAEL: Alligator.
BARBARA: Alligator? You want strangle and hit alligator?
MICHAEL: *(No response.)*
BARBARA: Who or what you want strangle and hit?
MICHAEL: Dog know.

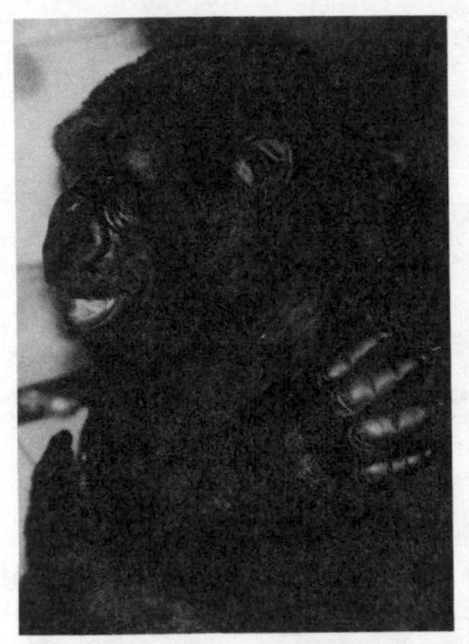

4. The Alligator Wars

Imaginary triumphs over possible alligators appear to be one of the ideas Michael holds dearest to his heart. In long and complex conversations, he sometimes appears to suggest that an alligator may bite Penny, Barbara, or someone else on the project—and he will defend them.

Does he always mean an alligator when he says alligator, or could he mean the coyote who attacked the lamb? (The sign for alligator is—literally—like snapping jaws.) Whatever picture he carries in his mind of them, alligators play a large role in Michael's fantasy life.

MICHAEL: Strangle alligator.
BARBARA: Who strangle alligator?
MICHAEL: Trouble Mike strangle alligator squash.
BARBARA: You did? You strangle alligator?
MICHAEL: Smile clap.
BARBARA: Oh. *(By voice—not a sign.)* You happy when strangle alligator?
MICHAEL: Squash.

5. The Red-Haired Rumpus

In the sign language that Koko and Michael have picked up from their surrogate mothers and each other, *lip* can mean what it seems to mean or it can be—and often is—used to mean *woman.* That clue explains something about Michael's report here on a red-haired woman's rumpus, described in Chapter Nine.

The dialogue is not a grammarian's dream, but it may have historical importance as one of the first recorded instances of a gorilla gossiping about the way the neighbors spat.

MICHAEL: Necklace. *(This means he wants the leash used when he goes for a walk.)*
BARBARA: Necklace?
MICHAEL: Girl. *(Michael stares out the window.)*
BARBARA: What makes you happy?
MICHAEL: Think necklace. *(A few minutes later.)* Know hit in mouth.
BARBARA: Know hit in mouth?
MICHAEL: Hit in mouth red bite.
BARBARA: Why are you signing hit in mouth?
MICHAEL: Know.
BARBARA: Who you want hit in mouth?
MICHAEL: Hair girl red.
BARBARA: What? Red-hair girl?
MICHAEL: Lips.
BARBARA: Lips?
MICHAEL: Lips lips lips big trouble.

Barbara concluded from this that Michael had been trying to tell her a "red-haired woman had been big trouble." She took the tale to the rest of the gorilla crew to see what might have happened while she was away.

Michael's bit of gossip was delivered on the day he had witnessed a lab assistant's angry red-haired girlfriend screaming and slamming doors in a towering rage.

THE CRITICS' CHARGES

An important part of the work at the Gorilla Foundation involves re-examining the films to make certain that what are claimed as findings are not in reality simply mimicry or the result of unconscious prompting on the part of the trainers. The arguments against the credibility of apetalk have tended to center on these two charges.

The charge of unconscious prompting, or cueing, has sparked a particularly lively attack on Penny Patterson and her confreres. This suggestion—that what the gorilla does is produced by cueing—is not as clear as it might be. Why is the gorilla so smart about which unconscious twitch to heed? There are also presumptions within this explanation for gorilla talk. If we put aside the notion of deliberate deception (and the critics do not claim that, though they sometimes hint), there is a presumption that (1) the gorilla can read signs so tiny the trainer is not aware of making them, and (2) the gorilla can read signs so tiny that none of the gorilla trainers find them when they reinspect the films and photographs. They do pick up, from the reinspections, obvious cases of imitativeness and promptings, but these are disallowed for purposes of the study.

The *it's-all-cueing* argument runs into another difficulty. The gorilla dialogues, at their clearest, are quirky. If they represent a *human* line of reasoning, humans are either strange or practical jokers.

"Elephant devil," "Think blow," "Hurry drapes"—these phrases attributed to the gorillas have a convincingly odd ring to them. The gorilla was taught *elephant* and it was taught *devil,* but *elephant devil* has certainly been presented as the gorilla's own idea of something fetching to say. While a maker of playlets *might* invent such phrases, we should grant to the recorders of the gorilla dialogues that the conversations are sufficiently chaotic and unexpected that the very tone of them can be treated as an argument for their legitimacy.

The thought of a gorilla talking is not easy to accept, but the thought of some postgraduates in psychology subconsciously cueing gorillas to sign "Think blow," "Hurry drapes," and "Gorilla me monkey" can be still harder to swallow.

Cueing is signaling. It is the suggestion: *Do now what we previously agreed upon.* A nudge of your husband's foot under the bridge table can mean, "I really mean it, we have to go home now." There is a great deal of adroitness in sophisticated cueing. It can certainly achieve miracles in two-party communication, concealed from a third person. We could even say: Cueing is a language in itself—gesture language in a sly or unconscious form. It is, nevertheless, quite different from mastery of a spoken or generalized sign language. Confusing it with the latter could easily cause trouble.

Think about the discovery Ingrid Kang made with the melonhead whales. With a wave of the hand, she could bring them to the top of the water. This is a sign, clear and aboveboard. But let's say this clever discovery, instead of belonging to a sharp-eyed, conscientious trainer like Ingrid Kang, winds up in the hands of a charlatan. If this charlatan trainer calls out to the whales, "Come out and nod your head if you know Jesus," a small wave of the hand might produce a seeming miracle. The whales would immediately burst from the water. Part of the crowd would go home with the impression that the whales have agreed they are born-again Christians. Now, no one happens to be indulging in this sort of trickery at the moment, but it is remotely within possibility, at least—through *conscious* cueing.

It is easy to see why critics want every circumstance of the talking-gorilla experiments carefully considered and why over and over they raise the question of whether the secret is in a system of cues—unconscious to the trainers, definite to the gorillas.

Cueing plays a meaty role in the objections to all apetalk. We will take a much closer look at this issue and its role in language learning in the following chapters as we see how the battle lines were formed and the opening shots exchanged between those who were certain that apes have learned to talk and those who are equally certain that they have not.

What are we to think, though, of a young woman who seemed to go so much further with apetalk than other investigators and claimed so much more?

Penny Patterson has favoring winds, I think, for the pursuit of animal questions for she can see the animal as a person, not an object; as an intelligence, not a dummy. From the beginning of her search, she was a believer, not just a searcher, and the belief has been useful in propelling the search.

Dispassion may be a useful quality in a scientist seeking to add to the record on a subject already well explored. But it is death to the explorer or inventor. To invent the car, Henry Ford had to believe that he could put parts together and that there would be a way to make these parts ignite into motion. Edison found his way to the electric light, despite endless discouragement, by his passionate conviction that it *must* be possible. With many such inventions behind him, he attributed his success to single-mindedness—the ability to exclude all else from his mind while he concentrated without letup on the goal he had visualized.

Penny was not doing a student's curriculum-inspired thesis on the pros and cons of gorilla talk. She was trying, like an obsessed explorer or inventor, to *discover* gorilla talk—to take all the bites she had to in order to make that discovery—and to be the first in the world to do so.

Researchers and investigators are everywhere about us. Discoverers are something else. They work from different rules—they are spurred by different passions. And those passions can be large. Dr. Francine Patterson, who is rather more than what developmental psychologists usually are, fits no tidy classification because she is not of the same dispassionate breed as those who conduct the great mass of behavioral experiments. The nature of her experiment insured that she *could not be* like others in the field, for only a person obsessed and inspired would conceive a notion on so storybook a scale as "talking with gorillas."

I heard a world-famed animal protectionist, a woman I admire quite as much as Penny, describe her as "Snow White and her big bad menace." But Disney's Snow White was a helpless sort led about by rabbits and fawns. She prevailed against the witch in the castle only because others helped her. Nobody can say that Penny has been protected, pampered, or led about. She carved her own path with incredible strength. She is, at a minimum, the gorilla handler *par excellence* of her own generation. If her experiments hold up and gorilla talk is as real as she presents it, she will rank as one of the profound discoverers in the last few hundred years of inquiry into the natural world.

11

The Dangerous Professors and the Struggle with Stanford

JACOB BRONOWSKI, CARL SAGAN'S IMMEDIATE PREDEcessor as a chief explainer of complex scientific ideas to a wide public, had a profound impact with his book and television series *The Ascent of Man*. Bronowski balked at accepting the new language experiments as constituting an ascent of the ape. In a 1970 *Science* magazine article by Bronowski and psycholinguist Ursula Bellugi, the authors opened the contention that while chimpanzees might indeed be "naming" things in a way that was new for nonhumans, they did not meet real tests for thoughtful, premeditated, creative language but were merely blurting out strings of words.

Bronowski and Bellugi by no means remained alone in their objections.

The idea that there could be an "ascent of the apes" brought about by human teachers in our own time would have been an amazing sequel to the story Bronowski had already told. But his opinion was unshakable. Bronowski, a scientific correlator of exceptional skill, found nothing to cause him to accept what he considered a cleverly disguised version of ancient fairytales.

By the late 1970s, when Jane Goodall began to send questions from the jungle in the hope that Penny Patterson's gorillas could provide some answers, it was clear to all who accepted the reality of apetalk that a transition was occurring between (1) simply talking to the apes and (2) taking useful information from them in the form of interrogation. Jane's earlier question on how one should approach an ape solved no mystery—she knew the answer before she sent in the question. But she had another question that neither she nor any human could have answered.

Jane wanted to know if a gorilla can tell individual humans by scent or if it merely identifies humans in general as having a particular scent. When this question arrived, the Patterson crew obligingly set up an experiment to try and obtain an answer. The characteristic smell of everyone in the trailer-house, humans and apes alike, was extracted from their armpits on swabs. Then the human crew members and the gorillas were tested for proficiency in determining which swab represented which creature at the Gorilla Foundation. The results would be meaningful only if someone could guess identities significantly above mathematical probability.

The winner: Michael. He was right in ten out of sixteen trials in identifying the smell with its source. Pure chance would have been 5.3 times. The humans and Koko scored much lower. Mike's answers seemed to indicate that a gorilla who understood the test had a better than human chance at knowing individuals by their smell.

Bronowski died in 1974. Would he have changed his mind about the ability of apes to talk if he had lived just a half dozen more years and been able to send his own questions in for the gorillas to answer, as Jane Goodall was doing? No one can say for certain, but the ten years following the publication of Bronowski's views on the subject have produced a vast body of documentation that was largely nonexistent at the time he raised a series of theoretical objections to the reasoning powers of apes. Bronowski argued that apes don't reflect, they don't conceive ideas about the future, and they can't visualize what is far off and personally removed from them. Where evidence was found that the ape *does* seem to have certain powers of conscious reasoning—as in the case of the slot-machine chimps—it was argued that a child of three is generally more advanced.

Yet, the feeling remains—and we will have a look at a series of brilliant expounders who argue this point—that progress in apetalk is not as great as the experimenters have suggested; that all the pushing and tugging at the primate intelligence has produced little evidence that the apes will actually *go somewhere* with language.

Have we, in this era of the Peter Principle, tried to do with apes and whales exactly what we so often do with ourselves—elevate them to one stage beyond their capacity? Was this what lay behind a vision of upwardly mobile apes who could learn human language or a dream of benevolent whales who could master our own tongue and then possibly teach *us* a thing or two?

Being what we are, with our all too human aspirations, this seemed entirely possible. Carl Sagan, though, understood the arguments of such linguists as Bellugi and naturalist-speculators as Bronowski and felt they had overplayed their hands. Sagan took the birth of an age of interspecies communication with great seriousness and profound hope, and he accepted apetalk as authentic. Although *The Dragons of Eden* grew from the talk he prepared as the first Jacob Bronowski Memorial Lecture in Natural Philosophy and begins with a salute to Bronowski's spirit, he went on to chide him, at a later point, for denying Washoe's significance. Bronowski and like-minded colleagues, he thought, showed "a little pinch of human chauvinism" in their determination to rebuke the experimental efforts.

To Carl Sagan, the evidence from the Yerkish and related experiments was too strong to be dismissed as just one more adventure in behaviorism that demonstrated human dexterity in manipulating animals to do new tasks. He invited us to believe that language learning, even at a rudimentary level, is not just a novel achievement—it is the key to a different life and to new societies. He asked us to believe that the apes have potentialities that are now being recognized and that the apetalk experiments can pave the way toward a more sensitive relationship between humans and the other animals.

THE CHOMSKY OBJECTION

Another set of objections was issued by Noam Chomsky, a linguist of formidable reputation then with the Massachusetts Institute of Technology. "Acquisition of even the barest rudiments of language is quite beyond the capacities of an otherwise intelligent ape," he asserted. In his own field, Chomsky has often been taken as the master, and many see him as both the soundest and most revolutionary of linguists. Acknowledging Chomsky, coping with Chomsky, revising Chomsky are among the foremost challenges for all who deal with the perils of language.

Chomsky argued that only humans have an innate capacity to learn language, as birds have an innate capacity for flying, because other species are not prepared, at the neocortex of the brain, to advance into creative speaking. Witheringly, Chomsky said, "It's about as likely that an ape will prove to have language ability as that there is an island somewhere with a species of flightless birds waiting for human beings to teach them to fly."

The dexterity of Chomsky's arguments appealed to many linguists. An ironist might retort, "Yes, because the consequence of believing Chomsky was to make the calling of the linguist supreme over all." His view resembled the old (and basically insoluble) argument over which comes first, the chicken or the egg? Descartes had suggested that humans, by developing language, could give momentum to their abstract thoughts and conceptions, building from idea to idea. Chomsky suggested that by having a unique ability for syntax built into their brains, humans were able to develop thoughts. Odd as the theory sounded, Chomsky felt that behavioral studies and brain studies confirmed his theory, and many agreed with him. But an ape who demonstrably talks would at least force a readjustment in the theory.

While Chomsky's wryest comments were directed at those who believed that apes were actually talking, the Chomsky Objection constitutes an indictment of the hopes for advanced animal communication in general. It could be interpreted as denying any possibility of conversation with porpoises and whales as well as with apes. Jacques Cousteau, as popular an explainer as Carl Sagan of ideas from the outermost reaches of science, had seemed to take his final position on the language question with this statement: "There is no obvious reason why man cannot learn the language of the dolphin." After a time, it became apparent that many who sided with Chomsky thought there *was* a reason why man could not: prevailing theory negated the possibility. Proponents of the

Chomsky position maintained that contrary evidence from the primate labs was somehow erroneous, and that further investigation would show that the experimenters had misinterpreted their results, or even faked them.

It was far easier for Cousteau, an adventurer as well as a scientist, to believe in marvels than it was for those who have not made undersea explorations. Although he did not claim to have spoken with porpoises, Cousteau seemed constantly on the alert for evidence that porpoises and whales are seeking to communicate with each other and with us. In a remark that can be heartening to those who believe in either whalespeak or apetalk, Cousteau had suggested that "scientists, in their quest for certitude and proof, tend to reject the marvelous." The ape experimenters believed that the Chomskyites had gone a step beyond merely rejecting the marvelous; they had also rejected the proofs that these marvels already existed.

The psycholinguist Jean Aitchison, an author and a teacher at the London School of Economics, retraced the ground of others in attempting to assess the authenticity of apetalk in her book *The Articulate Mammal*. After describing what had happened with Washoe, Sarah, and other chimps, she concluded that the results were definitive enough to weaken Chomsky's argument, and perhaps he would have to revise his dictum that even the most intelligent of apes cannot master "the barest rudiments of language." Even so, she didn't really side with the interspeciesists.

Defining psycholinguistics as "the study of language and the mind," Aitchison allowed that animals could be tugged into the use of words—the apes did it grudgingly, as all ape experimenters acknowledged. Yet her concession did not really discredit the essence of Chomsky's contention: that there was nothing truly creative about grudgingly produced words or even word combinations. Apetalk remained only proto-language, in Aitchison's view.

Chomsky called human speech "this still mysterious ability" and went on to assert that a *human* who failed to be creative in using language—someone who used only stock sentences and expressed only stock sentiments—would appear to be brain-damaged. "We would regard him as mentally defective, as being less human than animal," he said.

GORILLAS MAKE A CREATIVE SEARCH

The Chomsky Objection denies the possibility of a "talking ape" largely on the grounds that an ape is not truly a thinking animal and it can't be because it lacks syntax. We could turn the argument around and suggest that if gorillas are proven to be creative thinkers, there would be much justice in suspecting the language reports are perfectly valid. There are accounts of gorillas that seem to confirm that, in the wilds, they are considerably more intelligent—and their actions have a more thoughtful cast—than those of other animals.

In Rwanda in 1961, a small band of gorillas was wandering in search of a new leader. The gorilla naturalist George Schaller traced the activities of this group in his book, *The Year of the Gorilla*, and his description is reminiscent of tales of those families who headed west in America and then had trouble with the Indians. The gorilla band found a leader, but then a black leopard struck, killing the new leader as well as a female. Leaderless a second time, the gorillas went off in search of still another protector and, Schaller was told, they found one.

These were strange transactions under melodramatic circumstances. How could all this seeking and finding take place unless there had been complicated powwows among the animals concerned?

Without belaboring definitions of "the creative," a case could be made that these wanderings and the resultant treaties, as new leaders took up where others had left off, must have required something special in the way of negotiation and understanding. Schaller thought he could pick out at least twenty-one distinct vocalizations among the gorillas he watched and trailed. He made no case for "conversation," though, and the precise manner in which these tribal wanderings somehow produced new wagon masters for the group must be left to the imagination.

The family life of the wild gorilla is sufficiently developed that it sometimes appears to *require* a rather advanced use of language to be readily explained. Still, as we explore them, many of nature's greatest marvels—the migratory patterns of birds, the defense formations of the great herd animals—are found to exist not as a result of creativity but of compulsion. Nothing need be discussed in a town-meeting manner—nothing productive would come of such activity even in a gorilla tribe—if reactions to a given situation (*the black leopard strikes*) are dictated only by an inborn "gorillaness."

The idea of compulsion can be used to explain gorilla communication, but the explanation makes those "conversational gorillas" in Woodside more crucial than ever in the debate over animal communication. If you believed in the dialogues attributed to them, you glimpsed a creature possessed of a quality it is sometimes said that humans themselves do not have: free will.

A gorilla who invents his own slang and a gorilla who demands that the drapes be pulled when she thinks she sees trouble in the neighborhood blast a terrible hole in the Chomsky Objection if the feats attributed to them can be documented. Koko and Michael actually stormed the Chomsky Objection on two fronts: (1) their dialogues were well beyond "the barest rudiments of language," where Chomsky says the apes cannot go, and (2) the gorillas were creative in their use of words, which Chomsky says is an animal impossibility.

ESCAPING THE SKINNER BOX

There was a buoyant honeymoon period for the apetalk experiments in the early 1970s when the Gardner-Yerkish-Premack findings, despite occasional plaguing criticisms, were largely accepted. Then a backlash set in, and it's still going strong. A determined battle line of dangerous professors—*intellectually* dangerous, peers of their field—came forward to savage the experimental results by formulating theories that disallowed them, questioning the safeguards around the experiments, and limiting the definitions of language to exclude apetalk and animal talk in general as legitimate variants of "language."

Bronowski, Bellugi, Chomsky, Thomas Sebeok are among those dangerous professors. There are others. Perhaps the most formidable was Harvard's B. F. Skinner, the behaviorist who had invented many of the principles upon which even the ape communicators sometimes relied. The "Skinner Box," a device for manipulating animal behavior by offering rewards for the right or beneficial choice, was about as close to a household phrase as anything that behavioral psychology had produced. Skinner himself had shown that there is very little an animal cannot be taught to do. He had demonstrated that rats will master mazes and that pigeons can be taught Ping Pong, but only in rote steps. Skinner has always been involved in manipulating the instinctual gifts of animals, which can be very great, rather than bringing out their creative faculties.

The chimpanzees and gorillas were given rewards for good behavior, in the usual Skinnerian sense, to get them started using words. The experimenters claimed that when the language experiments had gone far enough, the semantic base that had been established through conditioned reflexes would provide a framework for back-and-forth discussions that would include choice and imagination. In other words, Skinner's methods were being used as a way of ultimately freeing these animals from the compulsions of Skinner's conditioned reflexes.

In the Skinnerian world, where a machinelike perfection is the goal, humans who believe they are freely choosing, creative thinkers can be troublesome enough. Should theorists have to deal as well with talking chimpanzees and gorillas, imagining themselves to have individual destiny and individual control of what they are saying, thinking, and doing?

The debate could verge on ultimate ideas about the nature of the universe and its creatures; but the attacks that left wounds came from scufflers like Sebeok, who openly challenged the experimenters' sagacity. The situation for the experimenters unquestionably became more grim when one of them—Herbert Terrace—defected to the other side.

A CHIMP WITH FIFTY-SEVEN TEACHERS

Dr. Herbert S. Terrace of Columbia University wanted to raise the chimpanzee Nim, whom we met earlier, "like a human child in a human family," but Nim's family became a multitude. We know just how many people worked with Nim, and for just how long, because Terrace, Harvard-educated and a student of B. F. Skinner, was the author of *An Introduction to Statistics*, and he never let a statistic get away from him. Neither checklister nor structural-physiologist, he took the same paths into the ape brain as the other communicators—but one day he announced that he simply hadn't found as much there as they had claimed.

Applying statistical analysis to the results delivered by Nim's mass of teachers, Terrace first felt appalled, then enlightened. Something was wrong with the ape experiments: Nim's performance was not like that claimed for Washoe and Lana. In Terrace's view, Nim was not a sentence maker, although he seemed to be; he was not an all-round creative talker, although he seemed to be; he was not, in the long run, a threat to the view that chimpanzees, however clever, now and forever lack the basic talents necessary for social or business conversation.

Videotapes allowed Terrace, who had been absent during much of the instruction, to review it later, after Nim had been packed up and sent to the Institute for Primate Studies in Norman, Oklahoma. The Columbia phase of Nim's life lasted from June 1, 1975, through February 13, 1977. It did not—as Terrace sometimes seemed to recall with relish—make history.

Terrace believed he could see glimmers from the chimpanzee mind, all right—he claimed that Nim seemed to regard him as "some sort of magician." There is a shattering description of Nim as an ordinarily affectionate soul who would indulge in "piercing screams for hours on end" when a favored teacher left.

Like other communicators, Terrace must have expected Nim would be a record-breaker when the study began. He even completed what he called a "tedious statistical analysis" and proved to himself that Nim, like other talking chimps, was a creator of sentences. But when Terrace sat down with the tapes, he had 19,000 utterances of two or more signs to work with, and his second application of video observation and statistical analysis toppled everything for him, enabling him to see his original errors. Nim was a terrible copycat and no grand creator at all. Terrace's report of this caused almost as much consternation as if he had just taught Nim to read the New Testament aloud.

In November, 1979, *Psychology Today* published Terrace's critical analysis of his own—and everybody else's—work. Although his book-length account, *Nim*, begins with a wonderful trumpeting about the excitement and profundity of the ape-language experiments, it also devotes a chapter to spelling out the bad news. For Terrace, this bad news was, in a way, good. His experiment with Nim had been humdrum; but now, by declaring that the Nim experiment "explained" all the others and showed them to be failures, he had found an almost magical way (living up to Nim's impression of him) of snatching importance from an experiment that didn't add up to the high hopes he had started with.

In the next chapter, we'll take a look at the methods Terrace used for showing others in error. Here we'll concern ourselves with his defection, which was a major event in itself. It gave heart to everyone who had ever attacked the ape language lessons on any grounds whatsoever.

Joining forces with Terrace in the pages of *Psychology Today* were Thomas Sebeok and his wife, Jean Umiker-Sebeok. Thomas Sebeok made the statement "Language is species-specific to man" the theme of his cause. Terrace found ways to straighten out his own thinking from the precise formulas of statistics. The Sebeoks took pleasure in their inside knowledge of the flimflam used with performing animals in circuses, carnivals, and on the vaudeville stage. They hoped to show that talking animals are a delusion—*just another animal act*—and Terrace was the angel of deliverance for bringing that message across.

The critics were satisfied that in the Terrace account they now had a way to utterly demolish the Yerkish-Premack-Patterson findings, but the other interspecies experimenters were understandably outraged that Terrace used *his* failure to condemn *their* work.

To Penny Patterson, among others, the circumstances Terrace had arranged for Nim's instruction did not remotely resemble the recipe that might be given for a successful venture into language teaching. She detected a great air of self-assurance in this Skinner-trained psychologist—accompanied by a woeful lack of knowledge as to what would constitute a fair trial of Nim's power to master language.

Conducting the experiment with the aid of several dozen students who happened to be passing through Columbia, where he was an extraordinarily busy professor of psychology, Terrace earned the enduring scorn of the other language experimenters for what they considered his misjudgment in constantly shifting Nim from teacher to teacher.

Only a handful of teachers have participated in the long saga of instructing Michael and Koko. (When confronted with a new teacher, Michael might go for a month with relatively little to say, apparently harking back to that old gorilla trait of retreating within himself.) In contrast, Terrace subjected Nim to an endless array of instructors in a three-year period. At the back of his book he names them all—fifty-seven, including himself. They range from a twelve-year-old, Fred Bever (a student at Cathedral School), to Dorothy Moscow, fifty-three. One of the teachers was Anna Michel, who, at the time she worked with Nim in 1977, was thirty.

While she is most careful to defer to Herb Terrace, claiming that she is not herself a scientist and that she did not go through masses of exhaustive data, as he did, to reach a conclusion, it is clear that Anna Michel came out of the Nim experiment with a different impression than Terrace did. She wrote a book for children, *The Story of Nim: The Chimp Who Learned Language*, on the experiences of the various teachers who worked with Nim.

During a telephone interview with Anna Michel I asked her if it was possible that Terrace, as a student of B. F. Skinner, had intended from the beginning to topple the evidence that apes were learning to talk. She replied that she was sure this wasn't the case. In fact, she could remember times when Terrace had grown angry if someone seemed to suggest that Nim wasn't doing well or that the language lessons weren't taking.

She knew that Terrace had always seemed to be busy, terribly busy. He was humorous—when he had the time. But there were teachers who were probably closer to Nim than Terrace himself; the rest of them, after all, weren't professors of psychology with a heavy schedule. "Herb Terrace," Anna Michel told me, "never had to get down to the nitty-gritty of changing diapers."

When I suggested that there was a bit of gossip around that perhaps Terrace had been stuck with a dumb chimp, that maybe this was the explanation for Terrace's stark view of Nim's sentence-making capacity, there was a long pause and then a hesitant snort.

"Oh! He was intelligent!" she said. "Nim? Oh! He could be very devious, Nim. Why, he wasn't dumb, he was very smart.

Herb didn't really have time—not as much time as he should have had to work with him."

One small detail from the life of Nim has always nagged at those who have tried to follow Terrace's battle with the other communicators. At the start, Nim had a longer name: Terrace had christened him Neam Chimpsky. He chose this name with Noam Chomsky, that pre-eminent philosophical nemesis of apetalk, in mind. Terrace wrote that he had done so because it would be quite an irony if Nim, after undergoing language lessons, should prove to be the walking refutation of the Chomsky theories on the limits of apetalk.

I asked Anna Michel again if there was any possibility that Terrace had foreseen the results he would proclaim, for some have wondered if the name was the tip-off that he really thought Chomsky was right. No, she said; she believed Terrace's explanation; he was a skeptic with a witty streak. She saw him as sincere when he believed in Nim's language ability, and just as sincere when he didn't. He had simply changed his mind.

Anna Michel thought it rather sad that the experimenters had managed to go to war on the interspecies issue. She said her most surprising discovery, from the assignment at Columbia, was the competitiveness of the scientists in the field. "Somehow, being scientists, I thought they wouldn't be," she said. "It was disillusioning." When I asked her if she thought the interspecies quest had become meaningful enough to make some great difference in the world, Anna Michel said, "I really don't think you can tell from Nim. The gorilla experiments are interesting."

LAST CHAPTER AT STANFORD

When she first launched the gorilla experiment, Penny Patterson was a doctoral candidate in developmental psychology. Her work with Koko and Michael earned her the degree but did not earn her the recognition and support of the institution that originally had nurtured her studies. To anyone who is not acquainted with Stanford, it would seem that the university would rush to recognize Penny's world-famous achievements by giving her a professorship and encouraging her to build a new generation of computer-oriented interspecies communicators on a scale that would make Stanford itself the center of the universe for this kind of study. Stanford had a rule, though, that the grad students can't turn around into professorships. The prophet has to be rebaptized elsewhere.

Penny Patterson and the gorilla crew were sent packing.

"Oh, I don't know where Penny Patterson is," a Stanford administrator told me. "I don't usually know where our graduate students are. They study, they leave . . ."

A coincidence in timing makes it appear that it was the Terrace counterattack which accounted for their sudden departure.

"That all happened about the time I was leaving," said Penny. "We knew we were leaving. And we knew that the university was going to great trouble to make sure that nobody, including me, was able to fight it through or even reach the person who could have staved it off. Terrace? Terrace might have given us one final shove."

On the day the Patterson crew departed, they left a marker behind. In an old, abandoned building, they had found a mock gravestone that said *Our Energy Lies Here*—something college kids had fashioned, part of an old protest. The creators of the Gorilla Foundation, which would survive and go on with its work without university support, inscribed the gravestone with these words:

Here Lies Academic Freedom
at Stanford

The Gorilla Foundation, started while the Patterson crew was still on campus, was established as one of those corporate fictions, a charity to raise money for the instruction of gorillas. Now it is the home where Michael and Koko live. This was not the freest of choices, but the ferocious determination of the gorilla crew pulled them through this crisis as it had all the others.

The exile from Stanford is, in a way, the most interesting of the Patterson episodes, and it is certainly the most crucial, although the experiment had been in trouble over and over. Originally, Penny was constantly in danger that the San Francisco Zoo, which owned Koko, would take back their charge. She used favorable newspaper coverage and any other means available to hold onto Koko. Trying to forestall a foreclosure on her gorilla (not a typical problem for a doctoral candidate), she raised $11,500 to purchase Koko once and for all. But the zoo said Koko should return to the zoo anyway, in order to be properly impregnated—a gorilla should not live as a nun. That sent Penny into second-stage fund-raising—for another gorilla. She stirred up an additional $12,500 in order to satisfy the argument by importing Michael from Austria. Bargains both—everyone always told her how cheap she was getting her gorillas.

Solving such practical problems has proved easier than coping with those professors who are eager to level her experiments with the eloquence of their theoretical arguments.

When I contacted Stanford administrators to discover why an experiment of this dimension had lost its place on campus—why the Patterson team was now on Skyline Boulevard at a poultry shack instead of enjoying the numerous perquisites of an advanced research team on the Stanford campus—this is the remark I heard: The Patterson experiment was "not found to be along the interests of any of the faculty members."

By now nearly every word that Koko or Michael "spoke" bristled with interest for a vast audience. What other inhabitants of apedom had given us such direct commentary as "Horse sad" or "You nut, not me"? Stanford's lofty explanation was clearly not telling the whole story.

One of the men I spoke with in search of further enlightenment was Professor Patrick Suppes (who is not a villain of our tale). He had been of great help to Penny when the gorilla experiments were still causing pleasant excitement in administrative corridors. Stanford is rich in supertechnicians who can quickly work miracles with computerware. At the Institute for Mathematical Studies in the Social Sciences, Suppes worked out a way for Koko to talk via machine. This promised for a time to highly complement the training in sign language that had given Koko her basic vocabulary, because computer data might end the disputes about Koko's signing power. Suppes provided an ingenious setup. Keys were color-patterned and categorized—parts of speech, words for feeling and action, a condensed version of the highly complicated English grammar. Koko punched a key, and the machine talked in response to her choices. They were trying to take Koko all the way into syntax, an effort still underway as I write this, but with the complication that the deal to use the computer on the original basis keeps falling through.

Computer use originally came free—that was one of the reasons the Stanford connection was important as more than just a place for the gorilla crew to hang their hats. Then, as the university's cooperation evaporated, a fee of $300 a month was charged. That may have been a pittance to the university's better-heeled clients, but to Penny it was a body blow.

When she was "kicked off" campus (her words), Penny was warned that she would have to prepare to lose her place on the computer. It was a matter of fairness to other customers—they had to have one minimum fee for everyone. Penny felt they were implying that what the Gorilla Foundation could afford to pay would always be too little. It was hard to understand this or even to be sure she had it right, but Penny did not give up. She is still fighting for gorillas' rights on the computer.

Karl Pribram was able to shed some significant light on why the university had gone cold on the gorilla effort. Alone among those brain scientists who had concerned themselves with the interspecies quest, he had been able to watch Penny from the very beginning, and had seen her advance from student status to an international figure, renowned for the work that started when he had gone with her to San Francisco Zoo to find an animal of manifest possibility.

It occurred to me that in spite of his role as the mentor who had brought Penny Patterson along, Pribram might have come to feel, like so many other brain scientists, that the gorilla quest had gone too far; that Penny was reading more and more into less and less; and that an experiment which struck many a magazine-oriented reader as the most notable interspecies experiment of all was not really inspired.

"Oh no! Oh, I wouldn't put Penny down as some others have done," said Karl Pribram. "I fought for Penny staying here on this campus. I fought for other things I thought could go on in the primate center. I fought and I lost. I could understand the other side. It does take an awful lot of money. And there was fear. Some reasonable fear, I suppose."

I asked him if he was surprised that a student had managed to bring off a world-famous experiment. With glowing gentleness, he replied, "I am never surprised when a student does something wonderful." He talked to me for a time about the money poured into government projects of all kinds, but how little there has been for the great quest into the nature of life on earth, the origins of human mentality, the deeper subjects that experiments like the apetalk efforts are aiming at.

Pribram mentioned the finger he had lost to Washoe. And, as though this had made him think of the answer to my question as to why Penny had been pushed off campus in spite of her own wishes, in spite of his, Karl Pribram said, "I wouldn't make a mystery of why the apes are no longer at Stanford. It's very simple. The university doesn't want to take the risks of handling large animals. I can't quite agree with that, and yet I see their point."

"And the decision rests with—?"

"The decision," he said firmly, "rests with a lot of people."

That could have been the end of it. It was a matter of large apes, then; apes who bite and bring on damage suits. But Dr. Pribram was not quite done with his revelations.

"Jane Goodall is one of the clues to what turned Stanford off. She had our students in Africa. And our students were kidnaped, held for fantastic ransom . . ."

The connection was almost complete. Penny herself knew other parts of the story—personal piques, run-ins with the campus martinets. But Stanford had not been led to abandon the gorilla experiments because it feared the clamor of theorists who were outraged that people were taking interspecies experimentation seriously. No, Stanford could have stomached a quarrel of that sort without pain. But it was learning that dealing with the great apes is not a simple matter; it is a risky business, in fact, an occupation filled with perils.

TAKING THE APES AWAY

On May 19, 1975, three Stanford students and a Dutch research assistant were kidnaped from the Gombe Stream Preserve in Tanzania, Africa. The news report said they had been taken by members of a guerrilla army based in eastern Zaire, across Lake Tanganyika from the area where Jane Goodall was coaching the students in observation of the wild chimpanzees. In retrospect, the humans bore more watching than the chimps; but we know the aggressive habits of humans, and there was still much to learn about jungle primates. Jane Goodall and her students were learning this, providing a healthy chunk of the total knowledge that we have even now.

In pursuing her unprecedented work, starting to nurture students who could greatly multiply what she was able to do herself, Jane ran afoul of the kind of ferocious humans who are as common on the shores of Lake Tanganyika as in the dense jungles of our Western cities.

Ned McKay, an assistant city editor at the *Peninsula Times*, a daily serving the Stanford area, remembers that the university's attitude on the ransom negotiations was to keep them hazy.

"Jane Goodall's title was co-principal investigator, and there was someone else there, a Stanford professor named David Hamburg." (Hamburg is now a professor at Harvard and was not answering my calls as to how he effected the coup of bringing off the ransom negotiations.) "The last student was not released," said McKay, "until July 25, 1975. They finally did get all four of them back. The actual terms, how they got them released, that was never disclosed.

"There was talk at the time that there was a fund drive. Exact nature of the settlement never disclosed—apparently as part of the agreement with the kidnapers. There was a brief news conference in which the students said they were all right.

"Well, the ransom took months of effort, and it was all done sort of behind scenes. Very little ever disclosed. Stanford wasn't saying how they got in touch with the guerrillas or how you manipulate out of that. One of those very, very sensitive situations."

"Hamburg was the one who found the ransom money and righted the situation, wasn't he?" I suggested.

"Well—Hamburg always declined to say any more about it. In the interim period after the students returned," McKay continued, "it was kind of recognized there was a risk in people going to Africa. As I remember, there was no real debate over whether you now stopped on this kind of program. It was agreed that it was a sensible thing not to take the chance. If one group could be kidnaped, others could be, too."

The story of the kidnaping was elusive, and had always been elusive, but the chill was smack dab in the middle of the Stanford administration's shoulder blades—big, cold, axelike, and never forgotten. Even though the students were safely returned, it was decided that there *is* such a thing as an experiment that is more exotic than a university's legal staff can tolerate. And so, on the whim of a band of guerrillas, something as outlandish yet chilling as that, the great forward motion of superpowered primate investigators like Jane Goodall and Penny Patterson was, for a moment, sent spinning. The incident affected Penny as much as it did Jane, for Stanford, having learned its lesson, took a new and dimmer view of experiments that were assuredly marvels but riddled with risk.

The control group of chimpanzees Jane kept at the university was disbanded and evacuated. Can a celebrated experiment of this kind be dispersed without acquiescence of the principal investigator? It could be and it was.

Penny calls Jane Goodall "my idol in good judgment, good behavior, intelligent thought." Jane needs her good judgment, for her whole life is an invitation to crisis.

"She felt responsible for the chimpanzees," Penny said. "She had brought them to Stanford. Without her, they would not have been there. And she was so helpless, because they were already gone by the time she found out. She heard—who knows what the truth was?—that half were being sent to New York for hepatitis research, and the other half to Texas for cancer research. She was in tears over what had been allowed to happen. She felt as I did—that the chimpanzees are persons, they're individuals, and that we have a moral responsibility because they are feeling, thinking creatures. They're our charges."

Who was to blame for the way the apes got kicked around on the Stanford campus? The guerrillas who took college kids prisoners were certainly partly to blame—they caused the university to get the wind up. But it is also possible to blame a certain lack of gumption on the part of the university authorities.

The dominance order at Stanford cannot be established in quick strokes as it is in the gorilla trailer-house. Penny found herself looking around for someone who could override everyone else. She was given to understand that if she could find a person on the faculty who would sponsor the continuation of her work as one of his great, compelling interests, she could stay on.

Penny says, "At the same time they were telling me—maybe not this straight—'But you'll never never find anybody.'" She found a willing faculty sponsor in the medical school. "But then they said, 'No, it has to be someone who has already done the work and is in the field.' *Who has done this work and is in the field?* I don't know what they said to Karl Pribram but I believe the intention was to scare him half to death. They told him he would have monetary liability. I think he wanted to support me, but the administration set terms so that he *couldn't* do it, so that no one would.

"They had ideas, visions, of gorillas costing millions of dollars. I said, 'We have liability insurance.' And then they said, 'No, even if you gave waivers, even if there were insurance, you can't really protect the university no matter how hard you try.'

"And the trouble with all this is: nothing has ever happened with the gorillas to indicate that there is any great risk. They were scaring people, they wanted *me* to be scared. It was suggested that I was in danger, that the gorillas could kill me. I would say, 'Why don't you come with me? Be in with Koko and see—see what it's like.' They never came. I'm talking about the administrative staff in the provost's office. They never came. They had their minds made up."

Penny was working her way to the top and tried to get there. All the academic opinion she had ever run into held that for an experiment on as great a scale as this—she is currently trying to build a half-million-dollar home for those ever-larger gorillas—she wouldn't be able to survive unless she had the backing of a major university. She received offers because of the publicity about the apes' language prowess, but the offers were from institutions located in the wrong climate. She wasn't going to subject Michael to another bout of the pneumonia he had contracted on the way from Austria.

She also wondered if it was true that great deeds collapse if a university says, "Ah, but it's too risky for us, dear girl."

It seemed to be the perfect break when Donald Kennedy became the provost, because he could be appealed to; Donald Kennedy, along with Richard Atkinson, had sponsored Penny's research at the very beginning. She believed that this boost in Kennedy's career was going to help her hold her ground despite those who might be warring to unload the whole chimpanzee-gorilla scene. She thought he would be "sympathetic and reasonable."

"I called," she said. "The staff said, no, I couldn't make an appointment; I couldn't see him. I tried calling several times. I tried to *insist* on an appointment. The secretary became really stiff. She said, 'I'll call you back.' She didn't, so I called again. She said, 'He doesn't want to see you. It's not appropriate.' "

The refusal to let Penny argue her case face to face with the university management seems, in retrospect, cavalier. Not that the fear of kidnap parties or the hazards in experimental work of the type carried out by Jane Goodall and Penny Patterson are in any way imaginary; experiments with the great apes, in the jungle or in the laboratory, with or without guerrillas nearby, are dangerous. Yet when was danger not the price for making advances that, a hundred years hence, could influence our perception of the world we live in? Audubon, pushing through the forests, was not playing it safe. Nor were the microbe hunters, risking they knew not what in the quiet of their laboratories.

It really *isn't* Stanford's obligation to provide a continuing supply of possible kidnap victims for guerrillas around Lake Tanganyika. The school has a right, perhaps a duty, to protect its students. But there was a second conclusion, down the line, they might have reached. Scientists willing to risk their lives need backing—all a great university can give. It isn't really a matter for lawyers to settle; it's the natural kind of risk a great school should be willing to take because within such risks lie the ultimate possibilities for the advancement of knowledge.

If Stanford flunked out in respect to Penny Patterson, its abandonment was by no means fatal. It didn't stop her from going on, and it seems doubtful now that anything can stop her except the loss of the gorillas to the diseases that always threaten them. Penny was thinking about talking with animals when she was a schoolgirl, before she had ever seen a gorilla. Trying to stop her by denying her space or program sponsorship or anything else would be like trying to deflect the Superchief with toothpicks.

12

War of the Worlds, Primatology Division

IN THE SPRING OF 1980, A SMALL BUT DETERMINED BAND of men answered what they may have felt was a sacred call to protect the public from being bilked by the animal communicators. They decided to hold a media event at New York's Roosevelt Hotel, complete with magicians and circus tricksters to show that the claims of the interspeciesists could be explained as either self-deception or deliberate trickery.

The ringmasters for what was widely described as a "circus" about interspecies communication included Thomas Sebeok and Herbert Terrace, whose credentials as professional kibitzers of the interspecies movement were becoming well known. Other featured attractions included a Zurich zoologist named Heini Hediger; Paul Buissac, an expert on animal deceptions of the big top; a magician calling himself "The Amazing Randi"; and a number of others whose presentations were aimed at confirming the notion that supposed feats of interspecies communication were, as Sebeok had been complaining, not science but animal acts.

The interspecies circus could not help but have a sparkle of interest. The idea seemed reminiscent of Harry Houdini's jousts, decades earlier, with the Boston Medium Margery and other spiritualists and psychics, whose ghostly goings-on he enjoyed exposing in mid-act. When he couldn't trap them in this way, he proceeded to duplicate their weirdest feats, explaining the magic he used to create a seemingly supernatural occurrence.

The debunkers of animal talk may not have had the dash of Houdini but they certainly had ambition. Although they were invited, most of the animal investigators at whom this event was aimed did not attend. Knowing Sebeok and Terrace, they suspected an ambush and stayed away.

Their suspicions were well founded, but their decision to ignore the ambushers was a mistake. It could have become a mind-tingling showdown between leading controversialists on both sides. As it was, the sideshow came off like an attack by a SWAT team on the clothesline of the opposing force. It's not hard to lick a clothesline, but it isn't the greatest of victories, either.

Sebeok's circus provides a useful takeoff point for dealing with the question of trickery or self-deception, though, because it became an entertaining parade of the high-level deceptions involved in "trained animal acts" and similar divertissements.

The event's featured speakers explained and demonstrated animal training methods for circus and stage with references to the secret "cueing" that tells the animal what the trainer expects.

One of these effects, called the "Kiss of Death," is spectacular. Roped to a bed, a beautiful woman squirms in voluptuous terror as a bear comes rampaging toward her. Then the bear is almost on top of his helpless victim; but what does he do? He calms her fears with a tender kiss. Quite a trick. The kiss of death is a kiss of love. The real reason for the bear's action, explained Paul Buissac, was its knowledge that the girl had a secret reward for him—she held a carrot in her mouth.

But there is more to this than the explanation indicates. The great circus trainers know that any sign of food can destroy the judgment of most larger predatory wild animals and make them dangerous to humans. If the bear "gently" takes a carrot from her mouth, instead of gently delivering a kiss, the bear understands very well indeed the nature of what is expected from him. *(The human stages a show and the animal understands his part.)* Only an animal with a lively mentality would be able to bring off this trick without harming the human.

At another stage of the interspecies sideshow, The Amazing (James) Randi, a professional magician who has sometimes duplicated the feats of Uri Geller to show that they are not truly supernatural, performed several tricks, including getting in and out of a rope tie at will. A pair of psychologists were instructed to keep Randi bound but couldn't do so.

"I mean to get across the point," the magician said, "that serious researchers who attempt to investigate so-called paranormal events and claims are considerably out of their depth." Randi's faith that psychologists aren't likely to detect a magician's mode of operation are justified. But the connection to those meticulous experiments with chimpanzees at the Yerkes Primate Center, or with other talking apes, seemed either slim or nonexistent.

To wrap up the affair at the Roosevelt, Sebeok appeared at a press conference the next day and, as usual, had the last word. "In my opinion," he said, "the alleged language experiments with apes divide into three groups: one, outright fraud; two, self-deception; three, those conducted by Terrace. The largest class by far is the middle one."

Nicholas Wade, a writer who covered the events at the interspecies circus, thought the suggestion that the experimenters were being accused of fraud was palpable—but unstated. He chided the perpetrators of the event for making large hints they were not prepared to substantiate. "Two psychologists who study deception, Robert Rosenthal and Paul Ekman, said that they expected the level of fraud in ape language research to be the same as in any other field of research," wrote Wade in the June, 1980, edition of *Science* magazine. "All refused to answer the question of whether they possessed any positive evidence of fraud by any of the handful of ape language researchers. A chimpanzee who asked not to be quoted by name told *Science* that among his species it is regarded as childish to make general accusations without supplying specific evidence, which the accused may then have the opportunity to refute."

OF WORMS, COMPUTERS, AND TERRACE ON THE PROWL

As the gaudy event at the Roosevelt indicates, the clash over apetalk is so naturally emotional, and so susceptible to dramatization by all parties, that it has provided steady amusement as well as enlightenment to those who merely wish to stand on the sidelines and observe.

Even at its most serious or most intellectual the debate tended to rage in colorful terms.

When Janet L. Mistler-Lachman and Roy Lachman, of the University of Houston's Department of Psychology, took up the question of the chimpanzee Lana's semantic competence in a 1974 issue of *Science*, they stated their opinion bluntly: "There is no evidence for semantic competence in Lana. What capacities can the authors reasonably conclude that Lana has? She can carry out nine or ten partly overlapping response sequences up to seven items long and discriminate those that terminate in reward from those that do not. Lana has definitely learned to perform longer and longer sequences for reward. Training animals to perform longer and longer sequences for rewards is not novel; it has been done with pigeons and even worms."

While a number of behaviorists have claimed that the linguistic performances so far displayed by porpoises or apes seem like more fanciful versions of the choice systems taught to birds, the Lachmans alone had thought of comparing Lana's much-advertised feats to the lowly worm.

Duane Rumbaugh and Timothy Gill reacted strongly to this charge, claiming that Lana often went well beyond the kind of rote response the Lachmans had referred to. "Lana," they replied in another issue of *Science*, "has come to ask for the names of items never before named, and she has then used the new names in sentences to request that the items be given to her."

Cleverly, the Lachmans suggested that if Lana were to be regarded as a nonhuman with linguistic capability, they could identify another nonhuman using an equally effective complement of words: the computer itself. Computers were being programmed, they suggested, to employ "not only some of man's language capacity, but also his abilities to reason, draw inferences, plan, and intend."

It is debatable whether any talking ape could be programmed as thoroughly as computers are often programmed, but there is another objection to this comparison. The computer is a human-created receptacle for exactly those procedures that humans choose to incorporate into its circuits. The chimpanzee is not created by humans and does not have an artificial brain. It is a free agent, at least until an electrode is introduced into its brain to influence its behavior by subnormal means.

When Lana cancels a grammatical error on the computer display console as she sometimes does, it is an independent act. If she feels lazy or distracted, she may not do it. Computers are never lazy or distracted; they may perform perfectly or malfunction, but no dollop of temperament ever enters into the process. Just as its temperament or mood can cause a chimpanzee to forget or misunderstand, there also exists the hope that temperament can lead the chimpanzee to expand beyond the lessons taught, to reason its way toward knowledge it hasn't been specifically given. In a small way—by choosing new word orders—Lana seems to do this, and larger ways may be coming.

The computer is locked to its instructions. A computer can say "Hug," and a computer could be designed that would *deliver* a hug. But a computer, given the instruction "Hug," would not think of replying, "What have you done for me lately?"

Nim, that chimpanzee Herbert Terrace taught, who has not been presented here as brightest in his class, had all sorts of attitudes toward the word *hug* alone. He could use it as a plea for mercy. Does a computer do *that?*

Terrace's criticisms of ape language ability were carried through on a larger, more detailed scale than the analysis of the Lachmans or those of anyone else. He was, remember, an expert on statistics and at times seemed to derive pleasure from the sheer quantification of it all. It may have been as natural for him to sift through 19,000 human-chimpanzee transactions as to line up fifty-seven teachers for Nim when three or four would have been better. By his own account, he made "unexpected discoveries" when he began to review the videotape record of Nim's supposed conversations. The basic discovery was that the chimp's expressions were, to a large degree, imitation of its teachers' signs. He seems to have arrived at his conclusions statistically after dividing the chimpanzee's utterances into five categories: *spontaneous, imitation, reduction, expansion,* and *novel.*

Once he had persuaded himself that Nim was performing no wonders in terms of creative sentence making, Terrace returned to the earliest claim for the discovery that apes can talk—he analyzed the Gardners' films of Washoe. He concluded that Washoe's observation about the doll in the cup ("Baby in baby in my drink," by Terrace's reading) was a coach-along job and that teacher Susan Nichols presumably hadn't realized the extent to which she was guiding Washoe through the signing.

Terrace reconstructed the scene in a sort of verbal slow motion. This is his play-by-play report: "Washoe is with her teacher Susan Nichols, who has a cup and a doll. Ms. Nichols points to a cup and signs *that.* Washoe signs *baby*. Ms. Nichols brings the cup and the doll closer to Washoe, allowing her to touch them, then slowly pulls them away, signing *that* and pointing to the cup. Washoe signs *in* and looks away. Ms. Nichols brings the cup and doll closer to Washoe again, who looks at the two objects once more and signs *baby.* Then, as Ms. Nichols brings the cup still closer, Washoe signs *in. That,* signs Ms. Nichols, and points to the cup. *My drink,* signs Washoe."

Terrace concluded that all he had seen was "a run-on" sequence with very little relationship among its parts. He added, "Only the last two signs were uttered without prompting on the part of the teacher." This last remark has such a parenthetic sound to it that Terrace almost succeeds in disguising what he was actually admitting—that Washoe had been dragged into saying the descriptive words for the doll but had suddenly grasped the proposition and made a point. For all the talk about mere imitativeness and cueing, the teacher's help was no greater for Washoe than that of a kindergarten teacher for a child who is learning to receive toys by name. By Terrace's own account, Washoe could identify the doll as *baby,* knew what *in* meant, and was positively possessive about *my drink.*

Terrace read the incident looking for negatives; the Gardners and nearly everyone else who saw the film, up until Terrace, watched it for positives; they *all* saw what they were looking for. But the Gardners have the edge in interpretation; Washoe was not blindly following the teacher's signs but instead produced the signs that fit the scene where her attention was being directed. If Susan had pointed to *that, that,* and *that* somewhere else in the room, an entirely different string of words would have been produced. This happens to be the essence of talking—in its learning stage.

If the family collie climbs into bed with a toddler and the child is given no more hint as to what he's supposed to say than *that, that, that* and can come up with "Dog in dog in my bed," we would be impressed and feel that baby is on the way to language. That's just how the Gardners felt about Washoe, and they had every right.

Terrace interrupts his account in *Nim* at various points to comment on the comparatively superior performance of children working the same kind of language puzzles given to the apes. If children weren't better at language than chimpanzees, Nim would have been the psychologist at Columbia and Herb Terrace would have been taking instruction on how to say "Baby in my drink." The point is not that children could do it better but that chimpanzees *can do it at all.* We had always thought they couldn't, and Washoe contradicted that. Nim contradicted it, too.

In his struggle to extract something forceful from the Nim experiments that hadn't been already reported from previous talking-ape experiments, Terrace became so intent on finding negatives that he couldn't sense the positives where they existed.

"During Nim's last year in New York only 10 percent of his videotaped utterances were spontaneous," Terrace wrote. "Approximately 40 percent were imitations or reductions. . . . To a much larger extent than a child's, Nim's utterances were variations of the signs contained in his teacher's prior utterance. He was much less likely than a child to add new information to a conversation in his replies." But when, before 1968 and the Gardner efforts, had any chimpanzee anywhere put 10 percent of its utterances into a human-invented word system?

Terrace describes "the emotional turmoil caused by the too frequent departure of teachers to whom [Nim] had become attached" as being greatly disruptive for Nim, but not disruptive enough to account for the difference between speaking child and speaking ape. By setting up the child as his standard, Terrace was certain to get a negative reading on the language learning of chimpanzees. He never fully departs from his practice of making this uninformative comparison.

Anna Michel's comment that she had found the scientists in the field all too competitive begins to sound strangely appropriate here in a way she may not have intended. Since Nim, with his plethora of teachers, proved no champion in language comprehension, it was as though Terrace, without guessing at his own motivations, had begun to look for some way in which the Nim experiment would reveal *more* than any other. Nim had set no records as a talker, so the chance for high achievement was in the other direction: Nim was the proof that chimpanzees *weren't* making the progress people said—*that was a discovery, too;* very newsy; Terrace rode it for all it was worth. It caused an ill-managed experiment to become just as famous as those which had surpassed it. But to arrive at this goal, Terrace had to work as hard in extolling the negatives of Project Nim as the Gardners had in extolling the positives of Project Washoe.

In the course of reanalyzing his results in order to get real value for time spent, Terrace misread the meaning of his own experiment just as he belittled the meaning of everyone else's. I don't think there was particular malice in this. By the time I had read the significant chapters of *Nim* a number of times, I began to have a hunch that Terrace was doing it for Nim as much as for himself. Although he characterized everyone else in the field as being hoodwinked by their apes, including most of the fifty-seven teachers who had worked with Nim, he made Nim himself a champion of sorts: the champion who dealt apetalk a body blow.

By concentrating on the down side of the Nim experiments, Terrace had a significant and marketable report, described by the editors of *Psychology Today* in the following words: "The future of these upwardly mobile primates seems to have suffered a setback. H. S. Terrace . . . and Thomas and Jean Sebeok, a team of anthropologists at the University of Indiana, argue that the apes appear to be able to learn words, but they cannot produce new and original sentences—the benchmark of language use. Without the ability to create sentences independent of intentional and unintentional signals from trainers, these primates cannot be said to be 'talking.' Most of those who work with nonhuman primates vehemently disagree with this conclusion, but many others feel that, at best, the pioneers in this research have overstated their subjects' abilities." Other journals put the gist of the Terrace counterattack still more bluntly ("Chimps No Use Language").

Toward the end of *Nim*, Terrace says that he was asked by the National Science Foundation if it would be worth "more than

$1,000,000 spent over a five-year period to advance the level of sign language in a chimpanzee beyond that demonstrated in Nim." He concludes that the spending of many millions "to produce a chimpanzee who truly understood the power of signs" would surely be worth it, and then goes on to an almost messianic message about the desirability of creating the kind of "community of chimpanzees" that Carl Sagan had seemed to predict in *The Dragons of Eden*. "It is hard to imagine a more exciting voyage back in time than such a community would provide," Terrace writes. "The opportunity to observe how the addition of language, as we know it, would influence the culture of a group of chimpanzees might provide a priceless glimpse of what life was like at the dawn of human civilization."

Since Penny Patterson's idea of acclimating apes to language included toting them about in a Datsun, it is doubtful that she was thinking of re-creating what language learning may have been like in primordial times.

There is a point in his book at which Terrace ceases to be just a gadfly for the communicators and becomes truly maddening. In his final projection ("Beyond Project Nim") of what lies ahead in the interspecies department, Terrace says: "Because I cannot overlook what Nim learned about sign language under conditions that were far from ideal, I feel confident that Nim's impressive achievements will not prove to be the last word."

To the interspeciesists whose efforts easily outclassed his own, a statement as bald as that from the person who had suggested that their own experiments could be mere bluff or balderdash was an intolerable whinny. It was as though someone who had just crashed a glider at Kitty Hawk, where the Wrights had already flown a plane, had then said, "I feel confident that my glider's impressive achievements will not prove to be the last word in air travel."

Once the crossfire started between Terrace and his targets, the apetalk argument soon shook down to fundamentals. In a snappish exchange of letters in the *New York Review of Books*, where Patterson, Jean Sebeok, and Terrace all had a go at fulminating against each other's line of reasoning, Terrace put his argument in its boldest form. While all sorts of memorable expressions had been credited to talking apes, Terrace said that something like this might be far more usual: *give orange me give eat orange me eat orange give me eat orange give me you.* Actually, as an examination of his book shows, this sixteen-word scramble is the longest single utterance he detected from Nim—and the point he makes about it, in the book, is that chimpanzee utterances stayed short, didn't grow, or else became a jumble. A fair point.

"The critical question," said Terrace, "is whether the ape is generating sentences or simply running on with its hands until it gets what it wants."

The gorilla dialogues answer the question. Most chimpanzee talk *does* tend to focus on demands—what the chimp wants now. This was certainly true of Nim, whose most typical expressions included *Nim me, banana me, more tickle, sorry hug, yogurt Nim eat, banana eat Nim, play me play, sweet Nim sweet,* and *grape eat Nim.* Since neither a banana nor a grape ever ate Nim, we can conclude not only that his mastery of syntax was incomplete, (so was the gorillas'), he did tend to gabble excessively in search of food and favors. The gorillas certainly do their share of gabbling, but often their dialogue takes on a different quality than this. It is not in *demands* or *wants;* it is a matter of observation, recollection, cryptic dreaming.

"Alligator bad frown eat girl," Michael would say.

When Penny asked, "What does *stink* mean to you?" Koko answered, "Koko know lip you know good. . . . Rotten devil know."

In their own tongue, half Ameslan and half gorilla, Michael and Koko produced a blend of thoughts whose meaning had to be fished for, but these meanings were quite different from a repetitive gabble intended to convey only an instinctual snort of hunger.

If gorillas become only *slightly* more intelligible they might, *bad frown,* eat all Terrace's theories at a gulp.

A MESSAGE FROM NIM IN THE 'MOST EXPRESSIVE' LANGUAGE

Terrace's neat maneuver of passing Nim off as the last word in language experimentation when he was really only the last word in a botched experiment made the other communicators wonder at the judgment of those who were paying so much attention to Terrace.

Anna Michel, trying *not* to sound like the last word herself, mentioned something to me that Jacob Bronowski and most critics after him had failed to note. It was an easy point to miss; you really had to learn Ameslan to appreciate it. She had learned, she said, that Ameslan is the most expressive of *all* languages, not excluding the spoken tongue she started in.

Terrace does not disagree with Anna Michel's assessment. "There is really no substitute," he wrote, "for seeing a signer convey all sorts of information simply by staking out various ideas in a signing space, by modulations of movement, eye gaze, facial expression, pauses, and other devices. . . . Sign language has its own poetry, slips of tongue (hands?), and puns."

Written translation, he says, cannot do justice to what has been expressed. Penny had tried to explain that all this operates in relation to gorillas, too, but no critic of apetalk was going to accept what the gorillas' teachers learned from eye gaze, facial expression, pauses, modulations.

After her own book on Nim came out, Anna Michel started on a new project—a book about a deaf person. I asked her how long it had taken the deaf person to master the sign language that had been used with Nim—a few days, a few weeks? Longer. A few months? "Over a year," Anna said. This carries more meaning than at first appears. If a human needs more than a year to learn sign language through spoken instructions she already understands, what a distance the gorilla has traveled to learn even the basic symbols of Ameslan!

After a time, Nim ran out of luck with Columbia, Terrace had to shut down the experiment, and Nim was sent to the Institute for Primate Studies in Oklahoma, where he was kept on a manmade island with other chimpanzees. He seemed to do well there—a chimp on the rise in the island's chimpanzee hierarchy.

After a year, Herbert Terrace went to visit him. According to Anna Michel, Nim remembered signs that he had been taught and greeted Terrace with "hoots of joy and signs for *hug* and *kiss* and *tickle*." Did Terrace cue him? Or did Nim just remember? If it reflected nothing more than a memory for simple signs, Nim wasn't doing so badly.

Terrace himself gives a more detailed account of this reunion. He says that Nim used nineteen different signs, including *shoe* and *out shoe,* which seemed to be an instruction to let him try on one of Terrace's shoes. (Nim had always been interested in covered feet.) This return engagement was covered by television. When Nim sat down with Terrace, he ripped open Terrace's shirt and discovered a wire connected to a miniature microphone that led to a transmitter in Terrace's pocket. It could seem almost creative of Nim to make that finding, but then chimpanzees are curious—always curious. Naturally nosy.

Terrace said that he had hoped to stay with Nim for a few days but his teaching schedule wouldn't let him, so he had to settle for a single morning. That had always been their trouble—there were so many demands on Terrace's time. To those who believe that an ape must be swamped in affection to make real progress in language, it is a fatal trouble; it kills the results that might have been.

Although Herbert Terrace made it appear that he had refuted Penny Patterson and others, his refutation was convincing only when he was the lone speaker and when the mathematics of his argument were accepted as though the math settled the matter. He wanted to talk in terms of two- and three-sign combinations and their absolute frequencies; other teachers of chimpanzees, like teachers of children, wanted to talk of those moments when, from meaningless jabber, something real, pointed, meaningful emerges. The Terrace analysis was quantitative, based mostly on Nim and on a review of the films and records of other investigators—a review that appears to me to have been quite as subjective as that of the original experimenter.

Terrace showed that Nim imitated and picked up signs from teachers, and that he was credited by them with saying more than he had said. That is unquestionably true.

It has been suggested that the other primate investigators read a great deal into the signing of their animals—that they were essentially "doting mothers" looking for the best in their babies' performances. That seems likely.

An experimenter *would* look for the best. But it is on that best—an ape's occasional breakthrough into intelligible thought, communicated through signing—that the progress of apetalk can be judged. To add up the ape's utterances and prove that most of them were of no significance is irrelevant. When I asked Anna Michel if Nim had from time to time made statements to her that were unmistakably clear, she replied, "Oh, sure—sure he did." Far-ranging philosophies were not Nim's style, but snatches of intelligible chatter run through both Terrace's book about a chimpanzee who didn't learn to talk and Anna Michel's book about a chimpanzee (the same one) who did. Anna's report that she and Nim *sometimes* made sense of each other's signs is of more value than Terrace's quantitative analysis for the same reason that someone who measures a baby's babbling and finds it to be 10 percent intelligible is correct in assuming that speech is on the way. The chimpanzee may improve that 10 percent only a few percentage points while the baby will improve it 90 percent, but this doesn't mean that we haven't seen the birth of apetalk; it just means there is still a long way to go. After eons in which a formal sign language was not known at all, we shouldn't be surprised to find that the apes' ability to absorb it, or our own skill in teaching it, is restricted. Perhaps we'll learn together how to increase the percentage of comprehensible talk.

Terrace's case is special, I think. With the evidence in front of him that Nim was using language in a perfectly significant way—*when Terrace went to the island and Nim managed to get his shoe off*—Terrace still held to his view that Nim, for all his cleverness, had to be led through the signings by teachers who tipped their hand, with cues, as to what they wanted. But Terrace hadn't cued Nim when Nim asked him to take off his shoe. Even if he did, Nim made the proper language association, talking about the shoe in words, and it was close enough to a completed thought that Terrace could tell, from the signs, what was wanted *(the animal speaks and the human understands).*

In his book, Terrace is having it both ways, which is why it is so easy for the unwarned reader to take *Nim* at first as an affirmation of the chimpanzee's ability to use language only to find out later that it isn't. Terrace wants to show that the chimpanzee was bright at that; if Nim couldn't get a sentence down in a form to suit the linguists, then it wasn't Nim's fault—it was the general condition of apes everywhere. This problem might be licked later, Terrace suggested, but so far nobody had achieved that. *Nobody had done better than Nim!*

That was really Terrace's point. *Nobody did better than Nim, no matter what they say!*

Terrace was a doting mother, too.

SKIRMISH OVER CUEING

Terrace went to more pains to best the interspeciesists, but Thomas Sebeok actually had the more difficult argument to confront. Just one long, reflective statement from the hands of a gorilla was enough to knock linguistical theory into a hat and polish off Terrace's argument (that the sentence-making power isn't there) at the same time. But Sebeok had settled on the same claim that had been giving animal communicators trouble for decades. *Even when you see it, don't believe it,* warned Sebeok, because *it has to be cueing.*

Despite the hints at fraud and connivance, he really meant *unconscious cueing.* The trainers directed their animals and didn't know they did so.

For several years now, Sebeok has been flaying the interspeciesists with what he calls "the Clever Hans principle." He grew so accustomed to describing the process of unconscious cueing in this way that he would describe any supposed achievement of the communicators as their way of "Clever Hansing" the public. What could he mean, and where did this come from?

Clever Hans was a horse in Germany at the turn of the century that seemed to be able to answer questions in arithmetic by tapping out numbers with its hoof. Hans belonged to a retired schoolteacher, Wilhelm Von Osten. Von Osten sincerely believed that his horse was a remarkable creature and didn't know that he himself was giving off the clues that led the horse to begin, and to stop, with the hoof tapping. Then a psychologist named Oskar Pfungst exploded the mystery—to everyone's surprise, including Von Osten's. He demonstrated that Von Osten was involuntarily jerking his head and that this was the entire control—the entire source—of the horse's actions.

Surely the many persons who had been fooled would notice all that head jerking by Von Osten? No, because Clever Hans really *was* clever, though not at arithmetic—he, and other horses, too, can detect head movements at one-fifth of a millimeter. The jerks were too small for Von Osten to notice, but large enough for Clever Hans.

To demonstrate how important this revelation was in the effort to develop counterintelligence against the interspecies movement, Sebeok had dedicated that bizarre interspecies circus at the Roosevelt Hotel to the ghost of Clever Hans.

Sebeok, when I contacted him long after the affair at the Roosevelt, was happy to take me through a more exact description of his suspicions on the high place held by cueing in the performance of the talking apes. I remembered that Nicholas Wade had been struck by the fact—he ended his article by referring to it—that there was another horse besides Clever Hans who did the same tricks, was never explained as satisfactorily, and could scarcely be accused of reading his master's head movements. This horse, Berto, was, after all, blind.

I asked Sebeok to take up Nicholas Wade's point about the horse who was blind but solved little puzzles anyway.

"Berto was one of the Elberfeld horses, a stallion," he said. (They have been called the "notorious" Elberfeld horses.) "There is some evidence that he was *not* completely blind. Secondly, it was shown that information could be transmitted by two different channels. A man is next to the horse with his hand on the flank. He outlines the question on the flank, tracing letters or numbers."

Even if the person were not Berto's master, Sebeok argued, the human would be apt to give off certain clues. "The pulse rate changes," he said. "Through throbbing of the pulse, a clue is given to the horse. Another explanation would be acoustically. The breathing rate of the person asking the question could give a clue."

Hmmmmmm. Cueing was very subtle indeed, then. If Sebeok was correct, a horse who couldn't do arithmetic was able to take strong signals from something as relatively undetectable as pulse rates and breath rates.

Although this makes the assimilation of cues seem quite as marvelous as speaking language, I have no intention of denying the reality of cueing, even in its very subtle forms. A few years earlier, when I had learned hypnotism in order to delve into the Bridey Murphy mysteries, I found that, once a subject had been hypnotized, I could take him in and out of trance by nothing more than the fractional blinking of an eyelid. In trainer-subject relationships, the tiniest of cues can be well understood. The prepared subject will fall in a trance from a cue. Will a prepared gorilla push the right computer symbol or make a sign in Ameslan on cue?

I had no intention of arguing with Thomas Sebeok the reality of cueing. It is entirely, provably real. My wife proves the reality of cueing to me often. It is her habit to detect love affairs unerringly by picking up the unconscious signals of the participants. The first time this came up she told me a twenty-two-year-old girl was sleeping with a certain fifty-year-old man. I knew both well. She did not know them at all. I explained that I had watched them working together for months and there was no sign of anything beyond a normal friendship. She said, "You missed it, then." As it happened, events were bringing their love affair into the open.

How could my wife, a perfect stranger to them, have known within the first few minutes she ever saw them? In a thousandth-of-a-second flash, which she can pick off as though it were frozen in time forever, she saw an indication by the girl of where the man's overcoat was to go. It is the kind of signal passed between intimates and no one else. The signal was much too low for me to catch. Cueing.

The person cued often has no idea that it has happened, even though he or she may proceed to act on the cue. Animals can have an advantage on us here if the horse can detect muscular movements that are ten times as minimal as the slightest signal that a human could actually pick up. Cueing is certainly communication, but then again, as Sebeok would say, it is not speech.

I asked him if he did not consider the reading of such codes as remarkable a communicating factor as speech itself.

"Why no," he said. "Nothing is involved but a signal to start or stop. It is still just *go* or *no go*." (Clever Hans would begin to tap the ground at the sign of *go* and stop at *no go*.)

Sebeok's tone said: *That's not much*. But I was not convinced.

I questioned Sebeok about studies which were showing that the vervet monkey has several different cries ready, depending on the kind of danger it is in. If an eagle appears, that produces one kind of cry. But a leopard produces a different cry. And a snake produces a third. If the swoop of an eagle produces one sort of outcry but a leopard something entirely different and the snake different still, then it sounds mightily like the monkeys are chattering something as direct as "Eagle!", "Snake!", or "Leopard!" But Sebeok finds it impossible to concede this. The whole idea appalls him. "Language," he repeats, "is species-specific to man." It turns out, on questioning him, that he does not deny to animals what humans roughly equate as "a word," but holds that animals are incapable of formulating "the sentence"—sounds that add up multiple observations and abstractions into one consolidated thought. And as far as the vervet monkeys go, he cannot agree that they are calling out *eagle, snake,* and *leopard.* What they are actually doing, he says, is merely locating the zone of danger. "I was surprised those studies made the news they did. This is not new."

But if different sounds are shrilled how can Sebeok, a professional linguist, deny that monkeys are making those fundamental distinctions that give substance to contrasting thoughts? "Help!" is a sentence meaning "My God, get me out of this!" When my wife uses it, "Snake!" is certainly a sentence. It means "I'll never forgive you." To a monkey, it probably means "Jump! It slithers," and then other monkeys look in the grass. But wherein does this fall short of language?

Sebeok has an answer to such questions. Depending on your position in the dispute about talking animals, this will thrill you for the finesse with which he bottles up an idea—or will exasperate you with his determination to grant the monkey so little.

"Danger up, danger down, danger sideways"—that's what Sebeok says the vervet monkeys are identifying when they shrill at the eagle (up), the snake (down), and the leopard (sideways). It is an ingenious answer. I don't believe it for a minute. Up, down, and sideways are more powerful abstractions than such solid items as eagles and snakes.

"I would say that the animal lives in space," Sebeok told me soberly. "Predation can come only from above, below, or sideways. It is natural they would have some signs to distinguish the direction from which danger is arriving."

What is *unnatural,* he thinks, is the idea that a monkey has a name for the danger. If you accept his contention that language is species-specific to man, you can see why he draws back at admitting that monkeys are able to separately identify three different enemies. For if they know eagle, snake, and leopard, who is to say they don't have other chatter—perhaps not as easy to identify—for each of the birds and animals among whom they live? For various plants and grasses? For the different foods they eat? For clouds in the sky and a storm on the horizon? To say that a monkey shouts "Eagle!" when it sees one is to envision the possibility of a language that may be unlimited.

If Sebeok is right, then a plane spraying defoliants should provoke the same cry as the appearance of an eagle ("Danger up!"). I wonder if that will prove to be the case.

Stop-start—go and *no-go—up, down, sideways.* These limited ideas, Sebeok contends, amount to signaling but do not amount to language, and for the aggressive gentlemen who conducted the event at the Roosevelt Hotel, it is important to contain anything that looks like "animal conversation" within those simple boundaries—explain them in those terms—or their case comes crashing down on them.

The interspecies claims gave Sebeok a fresh challenge after his many studies in communication technique—and he soon settled down to the notion that cueing is generally the answer. Studying the animal conjurers, Sebeok learned techniques that more innocent folk would never suspect.

An animal trainer who puts an amazing pack of performing dogs through their paces is generally accused of doing this through use of whistling sounds that the human ear can't hear but the dogs' ears can. But Sebeok explains: the trainer, if clever enough, doesn't need a whistle. There are trainers who could submit, like Houdini, to a complete body search, letting their private parts and the insides of their mouths all be checked for a secreted whistle. Yet they would produce their animal miracles anyway. How?

"A very common cueing device goes like this," said Sebeok. "The fingernail on the thumb is pushed against the fingernail on the index finger. The sound produced is inaudible to a human but perfectly audible to a dog. And there are even simpler devices. The trainer may be standing in front of the dogs, and if he wants to give a 'go' signal, he leans on the right foot. To give a 'stop' signal, he leans on the left. People often shift weight, so who will notice? Yet that simple act can be used to give the illusion of the supernatural."

Such subterfuge doesn't make a cad out of the vaudeville performer. If a rogue comes along who claims he has a mind-reading monkey who can tell the numbers on a dollar bill, the results can be flabbergasting, but they remain essentially harmless as long as it is kept in mind that monkeys aren't telepaths, that their trainers are simply tricky.

Penny Patterson, I know, regarded Sebeok as a twiddly man who pulled cueing out to explain anything at all that didn't match his own ideas on the limits of animal communication. Superbly tolerant of gorillas, and usually of people, her tone when Sebeok's name was mentioned was one of pure contempt. They had, as it turned out, a previous connection to each other—a connection that reminded me of Karl Pribram's statement, "I am never surprised when a student does something wonderful."

"Oh, Penny Patterson *believes* that Koko and Michael are talking back to her," Sebeok said. "She's extremely emotional."

"Have you seen her in action, detected something in the films?"

"She was my student. I was teaching at Stanford God knows how many years ago, and I think it was her first contact ever with animal behavior. I remember her term paper—it was a very good term paper. Had to do with killer whales, captive killer whales."

Sebeok claimed that most of the ape experimenters were surrendering ("The last holdouts are the Gardners and Patterson"), but that was not quite the case—I found no one who had actually given up. He seemed to nourish a hope that the circus might have vanquished these folk and he could go on to more pressing matters. "Oh, it's a side issue!" he exclaimed. "I have other things to do."

"When Penny was your student, were you on opposite sides, even then?"

"With Penny? No clashes! She was quiet. She was a charming, very quiet girl. And what happened after I left, I have no idea."

"And when Penny thinks that Koko is saying, 'Fine animal gorilla,' proclaiming itself, really—"

"It is fantastic to imagine," he interrupted, "that a gorilla should have a classification between various species. This is not believable. A gorilla would not distinguish between other categories. How could he tell what is human, what is a bird?"

"A gorilla wouldn't know a bird from something else?"

"The captive gorilla, brought up as a human being, would classify himself as *a human being*."

"Would the opposite then follow? That is, if you take those cases where a child has supposedly been raised by a pack of wolves and then comes back to civilization—do you think the human would see itself as a wolf?"

"But they are not true cases. There is not a single authenticated case of a wolf-child. Not a single one! Usually, these tales were of autistic children, pushed out by their parents."

"What would happen if a child did happen to be lost in the wilderness and could find only wolves for company?"

"If a human were in a wolf pack, it would probably be eaten."

We began to talk, then, about the howling of wolves. I offered him the thoughts of the interspeciesist Jim Nollman, whose adventures I will sketch as we go along, and told him how Jim believed that in the howling of wolves was an outcry that spoke of the wolves' outlook on life. That howling had seemed, to Jim, extremely complex and diversified—ever changing.

"No, no, it is naive to assume that wolves communicate by voice," said Sebeok, brushing this away. "It is well known they communicate by scent. Secondly, and very importantly, they communicate by visual means. The position of the tail is very, very important. Most people who are amateurs just assume that animals communicate by sound."

"They do howl a lot, so what would you think their motivation might be?"

"Of course, they don't do it without a reason. We don't always know the reason. In a singing-dog act, you play a certain pitch, and the dog simply repeats the sound. What the function of howling is in wolves is controversial. No proof. I think the wolves might use howling as a demarcation of territory. Many animals do that. The gibbon does it. It is very complicated behavior."

I have quoted Thomas Sebeok here so there would be some sense of his mind at work. It is a good mind, reveling in the bizarre; it is a repository of a great many obscure facts. And it is a skeptical mind. Skepticism, when it becomes a passion—and it has often been a passion—can lead to elaborate measures to enforce itself, to fight off the contrary evidence. It could even lead to something as elaborate as that specialized circus at the Roosevelt where all guns were blazing at the notion that there has been an actual breakthrough in teaching animals to talk.

Sebeok and associates proved a great deal that night. They exploded animal deceptions that are often misunderstood by gullible audiences. It is debatable, though, that they did justice to, or succeeded in refuting, the evidence for "the talking apes."

Let's now go the final steps with Sebeok's ideas on cueing and how it works. He had an idea about bloodhounds that, it seemed to me, could be set up as a puzzle. The puzzle would go like this.

THE BLOODHOUND QUIZ

Let's imagine that a master criminal has forgotten himself for a moment and landed in prison. Finding himself displeased with the amenities, he decides to escape. We will award him a convenient tunnel, which no one else knows about, for this purpose. Being a mastermind, he faces squarely the fact that the best he can really hope for is an hour's start on the sheriff's bloodhounds.

Now this arch-criminal has cohorts willing to send a helicopter to a concealed canyon twenty miles away. They refuse to risk themselves, though, by picking him up near the prison walls within range of the guards' machine guns. The first twenty miles are up to him.

Gad, it could be a disaster. Our mastermind sets to work on a plan that will throw pursuers off the trail. How exactly do you fool a bloodhound and its powerful sense of smell? Soon he concocts what seems to him a nifty scheme. First step: steal an extra set of socks, shoes, and business suit from the warden's office. Second step: supply his friends on the outside with a complete set of convict clothing. This convict outfit must be steeped in his own smell, so he wears it several days without bathing. He wants the bloodhounds to have plenty of scent to follow. Third step: have the outside friends drag this stinky suit down a path to the creek a mile beyond the prison and from there drag it on a long, twisting, meaningless trail. "I'd make a trail about two hundred miles long that leads right to the edge of a cliff. Let the dogs figure *that* out!"

Comes the night of the great escape. Naked but carrying the warden's suit in a nicely wrapped package, our criminal genius lightfoots it down to the creek, loses the real trail in the water, and eventually dons the warden's suit at a point two miles up where he comes out of the creek. He then enjoys one of the warden's best cigars and heads for the rendezvous with the helicopter at Chaisemu Canyon.

The sheriff, who catches all the warden's escapees for him, is a pretty clever cuss, too. At least he's smart enough to trust his dogs. It's his feeling that a master criminal would probably head straight for Chaisemu Canyon as a convenient place to hop a helicopter without being spotted. He has learned by experience, though, that the bloodhounds know best. Get them on leashes and let them pull you on whatever crazy trail they choose—that's the way to catch somebody. The dogs will go leaping through territory you wouldn't think a human could get through and yet, often as not, will corner their quarry.

"They don't call 'em bloodbounds for nothin'," he tells the warden. "Sometimes I think the blood itself is what they smell, 'cause I had a feller steal clothes off a clothesline and throw his own away, but they passed them discarded duds right by and still caught him!"

The escapee is just coming out of the stream when he first hears the howls of the bloodhounds back by the prison gates.

Question: Where will the bloodhounds look for him—and why? Can they track him through the creek? Will they wind up in the warden's closet sniffing the *warden's* clothes? Will the false trail deceive them? How good an escape plan is it?

Answer: Given all the circumstances presented above, the bloodhounds will probably head for Chaisemu Canyon on an even straighter course than the escaped criminal did; and, if the chase is vigorous enough, they will likely cut him off even before he gets there. The sheriff will chortle, as he has chortled before, "Them dogs is just dang geniuses, that's what they is!"

Why do the bloodhounds outwit our clever criminal's scheme?

Because the sheriff is holding the leashes: cueing. He knows where a smart convict would go, and the bloodhounds—who are certainly brilliant—react to the sheriff's touch as though his hands are a divining rod. What makes this story significant, from the standpoint of animal behaviorism, is the sheriff's innocence of the principal factor in making the capture: he doesn't know he is cueing. For a dozen years or more, he has been attributing miraculous feats to his bloodhounds and has never guessed that the dogs are not following a scent at all. They are following his own "secret hunches."

Bloodhound idolators will have to forgive me for this story, which is probably quite close to what often happens when search dogs go "on the trail." Sebeok explained, "The sheriff imagines that the bloodhounds have this fantastic sense of smell and can lead him to the hiding place, but what happens is just the opposite. Muscular twitches from his arm and hand are transmitted into the leash. He doesn't know he's doing it, but the dogs do."

THE MUSCLE-READERS

For initiates, deciphering a subliminal signal can be used to achieve miracles more amazing than anything presented during that night of exploded mysteries at Sebeok's circus at the Roosevelt Hotel. Books of magic teach the art of muscle-reading (not everyone can learn it). But muscle-reading is different from other magic. It is not a trick; it is a feat. At the end of his stage act, the professional magician Kreskin typically offers to allow a committee selected from the audience to hide some small object—a diamond, for instance—anywhere within a theater. Let's say that one night it is concealed in the mouth of a woman in a crowd of two thousand spectators. Kreskin, who has been secluded in another room, now races about the hall, hanging on to one end of a handkerchief with a member of the committee holding to the other end. He has promised that if he can't find the diamond his entire $10,000 fee for the night's performance will go to a charity. But Kreskin is not all that charitable.

After much excited running, while the audience whoops and wonders, Kreskin infallibly zeroes in—first on the right section of the audience, then on the right row, then on the right woman, finally on the right part of her body—and has her spit the diamond into his hand.

When a feat of this sort is accomplished by sheer dexterity and not by the planting of confederates, and it can be, it proves that humans' tactile sense is considerably more developed than most of us realize. Small but readable tremors are conveyed through the handkerchief or by a direct touch on the committee member's body. Essentially, the signals are the same *go* or *no-go* that we discussed earlier; *colder, warmer, you're hot, you're not.* Many an old-time muscle-reader used to astound folks by having the town's banker stand beside him while he fooled with the combination to the bank's big safe. Guided by the banker's involuntary movements, which the trickster could muscle-read as he worked the dials, this genuine safecracker would turn and re-turn the dial until the doors swung open.

Such spectacular applications of a common principle seem pure miracle when practiced by an adroit fellow like Kreskin. Still, every schoolboy has suspended a weight from a string to "tell the sex" of someone who puts his or her hand under the string, or to detect the sex of an unborn chick. With no conscious direction, the weight swings back and forth to indicate male, in a circle to indicate female; but if he were told that a circle meant male and that back and forth meant female, the weight would be just as obedient. Ouija boards are guided by the same kind of involuntary movements. Unknowing, the players typically accuse each other of manipulation. The planchette of the Ouija board, as it scoots about, will generally prove to be as good a speller as the players—and no better.

"All right," I can almost hear the Amazing Randi saying, "so if animals pick up this kind of stuff even easier than people do, then why can't you admit that this whole apetalk scenario is just a case of the apes givin' back what experimenters are cueing them to give? Who says the experimenters are liars? Maybe they just don't understand the possibilities."

But there, Randi, is the catch. They *understand* the possibilities. Animal work *is* cueing—up and down the line. To be successful with animals, you have to work close. Closeness *always* involves cueing, usually in many forms at once. The very meaning of a word comes to the baby because, as it is pronounced, the parent is using eyes, body, gesture, everything to convey what is wanted. "Have to potty? Have to potty?" Such first conversations occur in a literal frenzy of cueing. The parent uses everything in the world *to convey the meaning of words.* Isolated from such cueing—an even more basic form of language than words—no baby would ever learn to speak at all.

The basic criticism the communicators have of Terrace, whose own experiments did not succeed, was his failure to give his chimpanzee both continuity and intimacy. Because *language* is abstract, Terrace seemed to think the *teaching* of it could be abstract. But that's why centuries passed in which apetalk and whalespeak were not discovered. We never exposed ourselves to the fullness of the animals' communication system; animals were assumed—except by some eager pet owners—to have no chance of understanding ours. It is not that too much cueing occurs between humans and animals. The problem has always been we cued too little. How much do you suppose that smart horse Clever Hans could have learned if its owner had pursued the great sensitivity in its private reading of cues? Words are peripheral—thinking is central. But words coordinate thought, even for the thinker. Body language led to the other forms of language. But it is still an assist in learning.

What ape communicators found is that there's a point where the cueing and even the body movements once needed to convey meaning can be stopped—and yet, using words alone, the thought goes through, the word is learned, the system compounds, and one day a Lana, a Michael, a Koko is beyond cueing and able to relate words, and combinations of words, to what was originally grasped only from gestures and signals.

How do we know? What assurance would a Thomas Sebeok need that this is true, if he were willing to accept the evidence as it appeared?

At the Gorilla Foundation, Koko, waiting for that day when motherhood would almost surely come, could be found shaping the fingers of a favorite doll into the signs for *eat* and *more*. Dolls do not cue. And there was a significant incident with Michael, reported by a volunteer named Rhoda Mitchell, who had been working with him.

"What does hitting two fisted hands together mean?" Rhoda asked Penny.

"Chase," answered Penny.

Michael had signed this a number of times. When Rhoda failed to understand, he took her hands, deliberately formed her fists into the same sign, and gave her a push to convey what he wanted: a rambunctious game of *let's-chase-each-other.* The only cueing came from Michael himself.

That story of Michael reporting on the activities of a red-haired girl who went screeching and slamming around the primate center at Stanford is more than just a tale about a gorilla following in the footsteps of the humans around him and gossiping. Michael was reporting the events of the day to someone who hadn't been there. His report was close enough to reconstructing what had happened that Barbara was able to figure out that something stormy had occurred involving a red-haired woman. But Barbara couldn't have cued what she hadn't seen, didn't know about, and hadn't asked about. That's why learning that gorillas will gossip was a long leap forward. If a trainer extracts a gorilla dream about fighting off coyotes, the critic can say that the trainer is simply cueing to produce exotic-sounding words like "strangle" and "alligator" and then interpreting the results to suit. Nobody, after all, has watched the gorilla's dream except the gorilla. The trainers themselves accept the dream stories because they have learned that gorillas *do* bring in information from out of nowhere. Even though it may not be perfectly expressed, careful listening and many repetitions begin to disclose what the gorilla has in mind.

At a press conference on September 29, 1978, Penny Patterson made a report that has sometimes seemed to anger the critics, and inspire their disbelief, more than any other tale from the Gorilla Foundation.

"A new development is that Koko has shown a spontaneous ability for rhyming," Penny stated. "One evening last month I offered Koko a stalk of broccoli and she signed 'Flower stink fruit pink . . . fruit pink stink.' When I commented 'You're rhyming—

neat!' Koko's refrain was 'love meat sweet.' She has given rhyming replies to other spoken and signed words: in response to 'long'—'wrong,' 'blue'—'do,' 'squash'—'wash.' "

The critics jeeringly responded that signing is visual and rhyming is auditory, so a sign-making gorilla would never have a reason to rhyme. That overlooked the fact that while the gorillas' method of replying was in signs (or via the sounding computer), they were also able to derive meaning from the English spoken around the trailer-house.

Koko, said Penny, performed "a mental translation process to rhyme the English gloss of these signs with the English gloss of others, and then retranslate and output the gestural equivalent." This process is similar to what a poet does—the words are heard in the head and then written quietly on paper. Those who rushed to proclaim that a rhyme is only a rhyme when it clangs through the air like church bells had forgotten the form in which humans themselves put rhymes together—the sounds just clang in the brain.

"Every day," said Penny, concluding her press-conference statement, "I continue to be amused and amazed by what the gorillas do and say, but I feel we are only scratching the surface, only beginning to discover the extent and nature of the intellectual and linguistic capabilities of nonhuman species."

The trainers at the Gorilla Foundation were well able to note that moment of passage—and it was an evolutionary passage, at that—when the gorillas moved from merely repeating what was wanted by the trainer to volunteering thoughts and to trying, with a still limited power of speech, to unscramble a dream.

When I opened my first "wild west" books and discovered that fellow inhabitants of this world were not only Pete and Dickie Darling, who lived across the field and over the railroad tracks, but Indians in warbonnets who had lost their lands to invading white men, soon I could see myself as the white Indian chief who rode through the night, galloping like a demon to save his people from the pony soldiers. Remembering such fantasies, I didn't find it so astounding that, as a natural consequence of looking into magazines for the truths beyond him, Michael the gorilla began to dream and boast of how he would scatter the coyotes to protect his true family there in the trailer-house. The power of speech probably does not come without the power to dream and to daydream; if it did, it would be a puny power, hardly worth having.

Why draw back from the accumulating evidence that gorillas dream—if not in sleep, then surely in their waking lives? This is where the language lessons were always headed; it gives the lessons pertinence and makes them worth the teaching. All the gorilla bites Penny took, en route to this, become worth it.

At the point where cueing is becoming *less* important and words are taking hold, interspecies communication moves toward a new phase where words have a magic quality indeed—they snowball the act of thinking. The linguists are right beyond contradiction in the premise that words not only express thinking but facilitate it. We can construct a universe of mathematics from the numbers 1 to 10, but until these numbers were conferred—when we merely had the idea of enumeration or counting without the vehicle of "naming"—we were not yet ready to do even basic addition. Contrary to some semanticists, a human being without language can think; deaf-mutes were thinking creatures before the advent of the hand alphabet. Any gorilla family demonstrates that gorillas wandering in the wild can be even more clever than Koko and Michael in their trailer-house. A creature with language, though, has been given the power to travel *far from the original thought.* As Koko and Michael become older and wiser, we can imagine language taking them where no gorilla has gone before.

Sebeok's objections aren't trivial. But they don't reflect facts—and they stand in the way of moving on.

We know, by this time, that even the hardest-headed critics of the animal experiments do in fact attribute to animals a perfectly plausible mastery of the following concepts: *start* and *stop; go* and *no-go; up, down,* and *sideways.* They seek to argue that it really goes no further than that. When it is demonstrated that a particular animal may understand a short list of words, or even a long one, and that it may have a sense of categories such as color, the evidence sometimes seems strong enough that the critic concedes it but says there is still no sign the animal makes a sentence. When it is pointed out the animal makes deductions *(if . . . then)* in a somewhat humanoid form, it is replied that the connections between language and deductive ability have not been established. Logic could exist without language.

Such arguments may seem imposing in print, but they do not when measured against the steady record from the Yerkes Language Center, from the Gardners, Roger Fouts, Penny Patterson, a number of others. David Premack took a more strictly behavioral stance than others, but his experiments have been significant. Herb Terrace wanted it established that Nim had gone as far as other apes—or that *they* had gone no further than Nim.

That was a lost cause. Michael and Koko *were* in a different league than the Terrace ape, and so were most of the chimpanzees whose feats have been sketched in this book. So were a number of others we did not have room for.

Apetalk is here. The fact that Thomas Sebeok has been able to cause as much confusion as he has by insisting that the ape investigators are "Clever Hansing" people right and left would indicate that it is really the defenders of evolutionary-linguistic theory in its conventional modes who are looking for a magical explanation to outlaw a run of experiments that contradict their own earlier thinking on the subject. Chomsky doubts apetalk; apetalk *defies* Chomsky.

But why should experimental results *not defy?* And why should those who find difficulty squaring apetalk with their earlier ideas not return to the drawing board—like everyone else must do when new facts come in—and see where this leads them?

While the gorillas, as they stand, are "creative" in language use, they aren't *enormously* creative; even on the basis of the best examples given, they can't be seriously compared with a really hot-minded six-year-old. And it has yet to be shown that a second generation of talking gorillas or chimpanzees will seriously improve on the first. They may be stranded forever with a language ability that wouldn't get them past first grade.

Or, in a few dozen or a few hundred years, they may surprise us by living up to those glowing pictures Carl Sagan has painted of ape societies swinging upward with human help.

In the meantime, we can at least assure ourselves that there are animals who, newly equipped with the rudiments of language, will now and then come up with the right words at the right time.

In the midst of Sebeok's circus, which was supposed to point up the folly of believing in animal talk, Suzanne Chevalier-Skolnikov related how she had watched Penny Patterson have such bad luck getting Koko to do something right that Penny finally blurted, "Bad gorilla."

Koko replied by signing "Funny gorilla."

The conferees found it hard to believe that gorillas could ever be that quick with repartee. But sometimes they are.

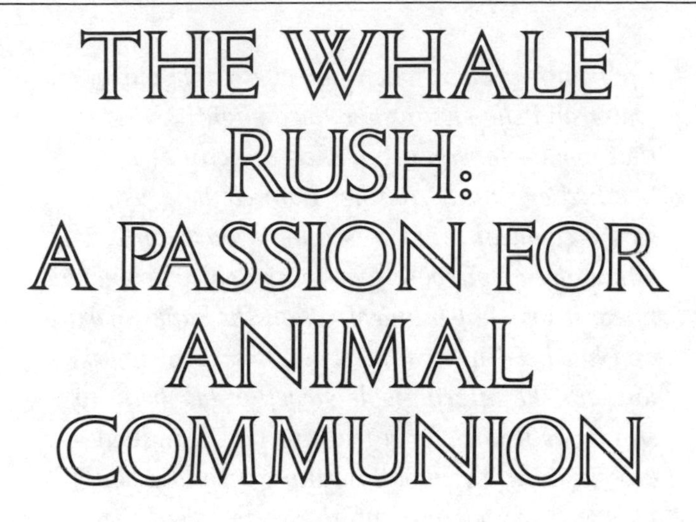

PART THREE

THE WHALE RUSH: A PASSION FOR ANIMAL COMMUNION

I felt no fear even when her eyes came out of water and she swung her head slightly so that one cyclopean orb looked directly at us. She had emerged from her own element as far as she could in order to see us in ours, and although her purpose was inscrutable, I somehow knew it was not inimical. Then she sank forward and her head went under. . . . It seemed to take as long for the interminable sweep of her body to slip by as it does for a train to pass a railway crossing. But so smoothly and gently did she pass that we felt no motion except when the vast flukes went under us and the dory bobbed a little. It was then I heard the voice of the Fin Whale for the third time. It was a long, low sonorous moan with unearthly overtones in a higher pitch. It was unbelievably weird and bore no affinity with any sound I have heard from any other living thing. It was a voice not of the world we know.

—Farley Mowat, *A Whale for the Killing,* 1972

13

The Receptive Humpbacks and a Siege of Orcas

JACQUES COUSTEAU HAD SAID, "NO SOONER DOES MAN discover intelligence than he tries to involve it in his own stupidity." It was fair warning, but it did not hold back the quest for whalespeak.

The great desire to break the code of the whales took a markedly different course from the experiments with chimpanzees and gorillas. It was different in technique, different in atmosphere, different in intention. Ape experimenters focused on definite, practical goals. Could a chimpanzee's vocabulary top a hundred words? Could a gorilla score above 70 in an IQ test? Advocates of the proposition that whales, along with humans, may be the mightiest talkers of the planet were primarily interested in deciphering the languages whales and dolphins are thought to use among themselves. *Learn their language—then you can speak with them.*

While the search for whalespeak has certainly been spurred along by evidence of successes with apetalk, the most ardent of the whale communicators sometimes expressed a feeling that when humans are truly on the right footing with whales, the humans will turn out to be the taught, not the teachers.

Although Dr. Lilly had started teaching words in a style similar to the chimpanzee experiments, he eventually abandoned this approach. Not before he deduced that the dolphin Elvar might be further along in sensing human meanings than Lilly himself was in sensing dolphin meanings. Lilly then became intensely caught up in the ability of the dolphins to convey attitude ("They're amused at us"), a common experience for others sensing a dolphin superconscious. Dolphin trainers of many persuasions have worked hard at reporting their transactions with creatures who may be quite as brainy as we are but are assuredly brainy in a different way, versed in a logic that (if it *is* logic) is quite unlike our own, possessed of physical powers that we do not share.

Confronted with dolphin test subjects who will shake their hands within a week of captivity and jump rope within eleven days, the trainers often stagger back, amazed at dolphin dexterity and the animal's easy grasp of human wishes. The presumption that there is something childlike in the test subject did not pass from the ape experimenters to the experimenters with cetaceans. Ingrid Kang, the porpoise trainer we met at the beginning of this book, explained that there were small whales who felt equal to herself in the dominance structure—and she accepted that.

The mood was different, then—but in one way the search for porpoisetalk and whalespeak was very much in tune with the search for apetalk. There has been a war of the worlds in the cetology division, too. Dr. Lilly—mocked but listened to, reviled but extremely influential—has been the target of most of the contention. It isn't as though he has been at bay before the mob, though. Sometimes the mob has been at bay before Dr. Lilly.

Understatement has not been the style in the porpoise wars any more than it was in the gorilla snipings. In summing up his objections to Lilly, Dr. Ken Norris, a professor at the University of California at Santa Cruz, was able to boil it right down. What Dr. Lilly had done, said Norris, was to "infect an entire generation with . . . poison." He told it to the *New York Times* so it wouldn't be missed.

ANOTHER WAY TO BRING OUT THE MAGIC

Ken Norris was the first man to capture a whale alive, and he was also the inventor of many of the training techniques and research methods used with captive porpoises. The author of an intense and engaging book, *The Porpoise Watcher*, which depicted his exploration of such newly discovered matters (in 1959) as the marine mammals' ability in echo-location (the porpoise sonar system), he came to be known as the greatest porpoise expert in the world. It fit him—if you didn't believe that this title really belonged to Dr. Lilly.

After the publication of his book, Norris was at times annoyed by what struck him as prejudice or naiveté on the part of some of the aspiring researchers in the field. He noted that he had arrived in his profession "like many another naturalist . . . via the route of the hunter and outdoorsman." Many latter-day whale or porpoise naturalists had arrived via Dr. Lilly and the interspecies movement. The young researchers Norris worried himself about were deeply hooked on the thought that humans must not only try to talk with porpoise and whale but try to accept their character traits as improving even for humans. Norris was a great believer in the power of the porpoise to work wonders. He just didn't believe in linguistic wonders. Since he wanted them to face up to the real, Norris suffered a bit of revulsion at the students who turned up at Santa Cruz seeming to believe that porpoises would prove to be perfect chatterboxes and that a kind of spiritual revelation might be achieved through a communion with these animals.

To Ken Norris, that was tripe of a low order. Dr. Lilly's soaring proclamations on the soon-to-be-fulfilled glories of porpoise-speak were bringing Norris a kind of student that he considered close to hopeless. Propagandized so thoroughly, how could they ever hope to get back to the truth of what whale and dolphin really amount to?

"You see it in applications for graduate school," Norris told the *Times*. "There's lots of this romantic view of life, all that kind of emotional stuff that happens when people of this cast of mind meet one of these animals. You hear about these peaceful creatures that live in a perfect society, blah, blah, blah, when in fact they can be as aggressive as hell."

One thing that Norris and others like him did not do was to underestimate Lilly's effect. Michael Parfit, writing about the dolphin dispute for *Smithsonian* magazine ("Are Dolphins Trying to Say Something or Is It All Much Ado About Nothing?") came across a Navy staff biologist at San Diego, Forrest G. Wood, who seemed to blame himself for all this Dr. Lilly trouble. He had made the mistake, he suggested, of introducing Lilly to the voices of dolphins on record tapes.

"I'm not sure if I hadn't played him the tapes the world would be a different place now," Wood told Parfit. "Nobody other than Lilly, no *scientist,* I believe, has attempted to say how intelligent these animals are with respect to humans."

At the Naval Ocean Systems Center in Hawaii, A. Earl Murchison, president of the International Marine Animal Trainers Association, told me he was personally friendly with Dr. Lilly—could sit down and have a great chat with him—but he was planning to suggest at a meeting of the trainers' group how they might counter unfortunate ideas that Lilly had put in the air. Dr. Lilly had caused the most difficulty, Murchison thought, by distracting attention from the dolphin's amazing ability to take sonar readings, to glimpse the inside of an object acoustically, to bring off feats no human could possibly accomplish—accomplishments unique to the dolphin.

Murchison wished the whole dispute could be settled, then left behind, so that people could learn "what the real magicalness of the dolphin is all about."

Professional trainers, researchers, and theorists can be so caught up in their own practical knowledge of porpoise and whale that they miss the reasons that this quest after interspecies communication is so absorbing and so desired.

The search for whalespeak has always been swept along by the belief, clung to in spite of everything, that there *is* an underlying grandness to the universe. Unlike apetalk, where the stakes are lower and the gorillas have merely been climbing toward mental status with bright young children, all those who feel a passion to find the talking whale can see something truly legendary in prospect—a chance to upend the world, a chance to correct old failings of the human race. This was the real pull of the quest after whalespeak—and no easy burden to put upon porpoise or whale, either. Even if the whales should reveal a language of their own, a complex social order, a set of traditions, a logic based on benevolence and gentleness, a talent for heroism.

SONAR CLICKS, A TALKING BUBBLE, AND INQUIRING HUMPBACKS

It is ironic that the very creatures the whalespeak faction deems to be the likeliest candidates for interspecies conversation are those whose communication system was virtually unknown until it was brought to light by accident during World War II.

The white whale, the beluga, was known to be a singer, and called the canary of the sea. Aristotle was aware of porpoise sounds, but this awareness seems to have eluded later humans. During World War II, while using a device called sofar (for "sound-fixing and ranging") to detect enemy submarines, the U.S. Navy coincidentally tuned in on that vast range of sounds produced by whales and porpoises. Each species tends to have its own distinctive sounds. The porpoise emits whole layers of sound. The fin whale produces an underwater sound that can travel a continental distance. At the outset of new and extensive efforts to understand the deep-ocean sounds, researchers were baffled by a sound so mighty they could not relate it to anything they had ever heard. They speculated that it might be the sound of the ocean itself, dashing against the continents. But they soon found this mystery sound was a call coming from the fin whale. To attract companions? To keep away enemies? We don't know the answer even now.

Intrigued rather than dismayed that there are underwater creatures whose traits even his own crafty brain cannot fathom, the graceful sea wanderer Jacques Cousteau has returned often in the past thirty-five years to the greatest puzzle of all: Do the highest of the marine mammals have some near equivalent to human speech? His answers are guarded but his examples vivid. The crew of his ship, *Calypso,* have often heard subtle combinations, not just instinctual yawpings. Cousteau has heard amazing concatenations of voices that seem quite meaningful even when they cannot be interpreted at all. In *Dolphins: The Undersea Discoveries of Jacques-Yves Cousteau,* he discusses whole fleets of dolphin clickings and points out that we don't even know what organ of the dolphin's body produces the sound.

"How can the dolphin make sounds without having vocal

cords?" he asks. "There has been much discussion of this and other questions, but no positive answers have yet been produced. For the past twenty-five years, thirty or so laboratories in various parts of the world have been engaged in breeding dolphins for study by perhaps a hundred specialists. Thus far, the only certitude we have is negative. We know that the dolphin's sonar signals do not come from its blowhole, and they are not accompanied by air bubbles."

Those clickings are usually described as dolphin range-finding equipment comparable to Navy sonar—an *inspiration* for Navy sonar, some would hold—and they enable the dolphin to send acoustical signals and read messages from the echoes. Cousteau separated the clickings from the other dolphin sounds and signals that he took to be their communication system. "Bubbles trailing like strings of crystal beads," he suggested, could be the sign of such communications, and he mentioned the claim that a single large bubble rising above a dolphin's head appears to constitute an undersea gesture of warning.

Carl Sagan repeated a theory that the mass of clickings coming from the porpoise might be a system for producing an audible "picture." The clicks, he suggested, could be a sound picture of something the porpoise apprehended (a shark, for instance) but with the picture expressed as a wave of clicks. "In this view," said Sagan, "a dolphin . . . transmits a set of clicks corresponding to the audio reflection spectrum it would obtain on irradiating a shark with sound waves in the dolphin's sonar mode."

Calypso engineer Eugene Lagorio would sit at the recording machine in a small boat, listening, in Cousteau's phrase, "like a wizard summoning up monsters." With his equipment, Lagorio could pick up the voices of as many as a hundred whales at a time, and he recorded what Cousteau heard as "roars, bellows, mews" or as "sounds which resemble trills, and the clanking of chains, and the squeaking of doors." These sounds didn't come from whales in general but from that single species of whale on whom much of the search for whalespeak has so far depended, the singing humpback.

One night the monsters emerged to nose around him—friendly humpbacks, freightcar-sized. With his headphones and his lights, dials, and general wizardry, he had stopped the whales on their planetary voyages. He heard the humpbacks make "little squeaking noises, like mice," and was certain they were having a conversation about him. "Maybe," he reflected, "they were wondering whether I was dangerous or not and if they should run away."

Is Lagorio attributing too much to the night sounds of whales? Cousteau himself knows that this is the danger.

"My friends and I have perhaps spent too much time with the whales," he says at last. "We may be the victims of an illusion. But how can we explain those alternating voices, and such a diversity of modulation, except by concluding that it is actually conversation?"

AN ABORTING WHALE WHO CALLED FOR HELP

Lagorio's visitors, the particular species of great whale that came to call, may or may not represent the best chance to find what interspecies communicators sometimes term an "extraterrestrial" speaking partner for humans. Humpbacks have been omnipresent throughout the discussions, though, as a result of the fame their singing habits achieved once we knew they sang at all.

The species deserves to be introduced on its own, for it often dominates the discussions of what the prospects are for whale conversations.

Humpback, an ungainly name if there ever was one, was the whalers' name for a creature that divers in Hawaii began to call "The Great Winged Angel." The difference in the two names speaks loudly of the great change that has come about in attitudes toward whales in those few short years, marking the end of American commercial whaling and the beginning of the search for underwater communion with the great whales.

America was done with the whaling in 1971. A quest after whalespeak had begun a few years earlier but was not transformed into a large-scale cult movement until early in the 1970s.

Porpoises, as friendly as they seem in aquariums or speeding at the bow of racing ships, have proved elusive to skindivers.

The great whales, though they *seem* remote and unapproachable, are surprisingly charitable to skindivers, allowing them to prowl about their great bodies. The humpback is most receptive of all.

Whales often react to the shouts and calls of children in rubber rafts. Other animals are apt to flee at such sounds. The humpback inspects. It is almost as though the whale is concluding, "The creature makes sounds. Possibly this is only instinctual, but suppose this is speech of some sort. These may be intelligent, reasoning creatures . . ."

Whether these meanderings of whales toward shouting youngsters have tangible meaning or not, there are various signs of the whale's high intelligence. Its mating games are astonishingly complex, and, seemingly out of sheer high spirits, the humpback may indulge in playful leaps through the air—a thirty-five-ton lift whenever it leaves the water. California psychiatrist Dr. Sterling Bunnel, in an article entitled "The Evolution of Cetacean Intelligence," offered a view that many in the interspecies search remembered: "Despite its low status in puritanical value systems, play is a hallmark of intelligence and is indispensable for creativity and flexibility. Its marked development in cetaceans makes it likely that they will frolic with their minds as much as with their bodies."

What happened to the humpback above the water, though, seemed minor compared to the figure it cut below. The divers arrived at that name Great Winged Angel as a translation from the Latin (*Megaptera novaeangliae*); they were moved to use it because of the whale's winged appearance when they encountered it under water.

Strikingly supple, a creature whose athletic grace the divers could not have imagined merely from seeing its hump at the top of the sea, the Great Winged Angel was a momentous experience for anyone with the nerve, luck, or chutzpah to attempt a rendezvous. Such meetings had to be carefully planned, for the divers must dive into the whales' path. Humpbacks travel fast at the top of the water, and, unless they choose to stop for a look, the diver will be left miles behind in moments.

How did the skindivers dare to encounter these enormous creatures? When did it start and what was intended?

The organized part of the whale swims, started in the early 1970s by divers who felt there could be a scientific advantage in the swim, was carried out in humpback nursery grounds at Hawaii. The excitement of finding the humpbacks often friendly caused such commotion that soon the area was being highly exploited by tourist boats and flanks of the inquisitive. "Whales can't go slumming," a diver complained, "so the slums came whaling."

A full-grown humpback can be fifty-three feet, and its "wings"—two powerful flippers, each fourteen feet long—are potentially lethal weapons. The Great Winged Angel treated the tiny creatures from the upper earth with an idle, curious benevolence. Although whales have the power to kill, they do not exercise it—or almost never have—on humans.

In March of 1976, not long after the discovery that a diver does not have to fear for his life when approaching a humpback in the sea, an incident occurred that is certainly the apotheosis of all the tales about communicating whales and humans who think they can read the message. The incident is reported in greater detail by Roy Nickerson in his book *Brother Whale* about his whale-watching days off Hawaii. This is the story:

One of the giant humpbacks knocked its head on the tour boat *White Bird,* traveling between Lahaina and Lanai. "The whale repeated the head-bumping three or four times," reports Nickerson, "each time backing off and directing her giant head upward to look at those on deck." Believing that the whale must be asking for help, and that this was the only possible explanation for her behavior, diver Randy Coon, whose parents owned the tour boat, dived down and discovered that the whale was in the midst of an aborted birth. Unable to help, he returned to the boat, but later, according to Nickerson, other divers went to the aid of the whale, using a lasso to pull the dead infant free of her body. Thus they gave the whale a chance to go on with her life.

Technically, however, the divers had broken a law intended to protect whales from molestation. Professing not to know the names of those who used the lasso, Nickerson nevertheless claims this as a genuine incident. If we accept only the first part of the tale, which all the passengers on the *White Bird* witnessed, it still seems a remarkably complete communication pattern—the whale knocks until it is understood, the human goes down to take a look. If all is true, then the whale has been very smart, the humans very daring, the whale very understanding—as they remove a baby half in and half out of her body.

The whale's persistence in knocking and signaling until she was understood appeared to demonstrate a different order of intelligence than humans were used to in nonhuman species. Dr. Lilly's speculations on what he called "the great computer" (the brain of the whale) appeals to anyone who believes that there is continuous evidence of whales *thinking* their way, not just *instinctualizing* their way, through a whale's daily grind. In the 1974 anthology which seemed to sum it all up, *The Mind in the Water*, editor Joan McIntyre stated the principal paradox: "It is very hard for us to accept that the actions of the monster might be specific. Or to relate to the intricate and detailed way a mountain of flesh and bone can act. But it may be that our difficulty has as much to do with our own bias and culture as it does with the creature we wish to understand. We expect that anything that big would of course be wild and random."

Project Jonah, a worldwide whale-rescue movement that Joan McIntyre headed as president, promoted a spiraling interest in the mental and verbal powers of whales. Stunning reports of the mental acuity of the great whales rolled from the gaily named Yolla Bolly Press of Sausalito, California. Across the Golden Gate Bridge from San Francisco, Sausalito was packed with the stormiest pro-whale contingents in the country. The more stalwart of those who agreed with Project Jonah that the whale is a creature of surpassing intellect sailed with members of Greenpeace to search out those far-ranging Russian or Japanese whaling ships and try to intervene between harpoonist and whale. Out of these Greenpeace confrontations came one amazing fact about the tactics of whales under attack. No matter how often the Greenpeace crews maneuvered near the whales, they somehow came through unscathed. Even though a harpoon could fly over their heads and strike the whale, they were not smashed to oblivion by the threshing flukes. Even in its death throes, a whale seemed to avoid any wild wallops to those who had sought to protect them from the harpoonists. Did the whales recognize that the Greenpeace valiants were not their enemies? Whether they did or not, they never hurtled them into the sea, even though the whales' last chaotic strivings—with life ebbing away in titanic jerks—could be terrifying.

Even in their death throes, the whales appeared to be thinking, benevolent creatures.

MUSICAL WHALES AND WHALES IN THE STORM

Roger Payne, a neurophysiologist Ph.D. from Cornell, began his research in animal behavior by studying how an owl can find its prey. When he turned to the study of whale songs, he became world-famous. His record album, *Songs of the Humpback Whale,* was a surprise hit. He selected the songs from a vast number of recordings made at sea. Inspired by Payne's extractions from whale musicology, several composers have now written orchestral works for "whale and symphony." Record reviewers have treated the humpback whale as a welcome personality amid the standard country-pop virtuosos and pretenders. Reviewer Peter Marshall claimed that the whale serenades, as recorded by Payne, had "freed folk-acid-rock from its pedestrian rhythms." Spinoffs, now multiplying by the hour, have included *The Songs of the Wolves* ("a great gift for you and your dog"), *Sounds of North American Frogs,* and *Sounds of a Tropical Rain Forest,* featuring howler monkeys and the three-wattled bell bird.

The measured way in which even Payne's most advanced conclusions were put before the scientific community becomes evident in a 1971 discussion presented jointly with Douglas Webb of the Woods Hole Oceanographic Institution. This paper took up the sonic range of humpback vocalizations. It was explained that the word *song,* as familiarly used, can be fastened on most any "repetitive, patterned sounds" from birds, frogs, insects, and the like. "Although the function of the [humpbacks'] songs is unknown," the authors wrote, "they have impressed many human hearers with their surprising complexity, and many people seem quick to want to ascribe some advanced communicatory function to them. It is well to remember, however, when trying to assess their function, that these sounds are within fairly narrow limits, monotonous, and anyone who advances a theory that is to explain the songs satisfactorily cannot afford to overlook this fact."

In other words, *no shocks here.* Having made the going seem so safe, Payne went further. But he never did lose the knack of seeming to explain away large conclusions. It is a symptom of his devotion to tangible data, rather than psychological deduction, that his strongest statement on whale hearing is simply that he finds "strong evidence" that whales "hear best at those same frequencies at which they speak loudest."

Well, Roger Payne takes absolutely nothing for granted.

And yet: the particularity of his observations, the great finesse and comprehensiveness of his studies of whale sound, show that he is sensitive and alert to the small signs, after all. Not the subliminal signs; those that are tangible. He looks for them; he records them. As methodically as he has done this, it has made great waves.

Through the 1970s, Payne continued to make the most inventive contributions to the rigid, mathematiclike studies of whale sound. The Great Winged Angel has retained its title as the finest cetacean singer, such slurs as "monotonous" drifting away on the wind. The humpback could be heard with its song in the spring; and then, the next spring, as Payne was ready and able to demonstrate, the song would be changed. Evolved.

Since love was the whales' main business during the singing period, it seems likely that the most insistent part of what the whale sang about, therefore the part that should yield most easily to translation, will prove to be about desire, intercourse. This is what a great deal of other chirping and moaning, including mankind's, is about. A high percentage of our popular songs are the record of our desire, our frustration, our persistence in love, and the way the lovers of 1970 see it compared, for instance, to lovers of 1911. "Why Don't We Do It in the Road" springs from the same general source as "A Bicycle Built for Two," but there's a nuance of change there that is not to be missed. The whales came back, from their global treks, with the song a bit changed? So do we, even when we don't go around the globe.

Nowhere were the whales' high consciousness and special reasoning powers better displayed than in the hijinks and escape methods applied to love play. The naturalist is not just playing up to prurient interest in recording these matters. Not as self-conscious as humans (they don't have to hide their lovemaking away—indeed, they couldn't) the whales seem far from the purely instinctual level in their dalliance. Love is on their minds, and the naturalist in Roger Payne takes heed of that. This is his *National Geographic* description of males seeking, and a female dodging, intercourse among the right whales of Patagonia: Resisting advances, the female would go to the surface belly up. But she couldn't breathe that way, and the males knew it, so they would swim in circles, waiting for her to turn over for a quick breath. "When she [did] so," reported Payne, "the males quickly [dove], pushing and shoving to be the first to get into the proper alignment for mating." (Men who think their lady lovers ask rather too much of them should consider what the whale must do. The male's job is to get underneath and hold his breath while the lady takes her ecstasy at the surface).

Females who did not escape the pack with a belly-up position had another escape—they could dabble in the shallows, temptingly, in spots where there was no room for a lover to get underneath. But this was not as neat or frustrating a trick as the maneuvers of a lady whale the Paynes called Troff, familiar to them for the way she evaded sex. Troff's trick was to put her tail high in the air so the male would do the same—and work himself into a physically impossible position. If this doesn't sound too hard, try to have sex at an ice-skating rink—standing on your hands. To make her getaway, Troff would "revolve slowly," and another male would be beaten in the mating game.

Troff must have been a popular lady, for all her tricks, or the Paynes wouldn't have seen her with her tail in the air so often.

Do the whales have "Hard-Hearted Hannah" songs about ladies like Troff? We don't know that they do, but we should not imagine that whale character doesn't run a gamut—that whale history does not divide into epochs. If Napoleon speaks to the spirit of one century, and Gandhi to another; if Emily Dickinson is succeeded, in another century, by a girl singing about a "Heart of Glass," why, who are we to think the whales are less than this, less *changeable* than this? Why do we see all creatures save ourselves as psychologically nonevolving? The changing song of the humpback may not be translated, but it tells us something we needed to know: *this* year the whales are not the same as last year.

Once we talk, we will also know, as distinctly as we know from Troff, they are not the same as each other.

Great stacks of whale recordings, from Payne's expeditions and others, give us a "data base," now, on several species of whale. But even after that data base is ten times what it is today, there will still come a point when the scientists' point of view falls short and the final interpreter will be the artist. Our data base on Napoleon is, after all, the greatest for any human in history, but he existed as a whirlwind figure of a thousand possible motives, casting spells on all he touched, and the data base does not contain the emperor's soul.

We have been looking at Roger Payne the scientist, a cautious man. To complete the picture, let us look at Roger Payne the artist, casting off that cloak of caution and sensing something, out there in the wilds, discoverable only by allowing his spirit to go beyond the precisely recorded facts.

Besides all those whale charts, it was Roger Payne who gave us a picture of the whale in the storm, showing that he was far more than just a tinkerer with the new acoustical technology.

In that great cape off Argentina's southern tip, the real giant of the place is the wind itself—a wind so constant and stormy that the whale studies Payne was making always had to defer to what the wind was doing in that particular moment. Right whales of Patagonia, though, do not defer to the wind. Payne, looking out on this on days too miserable for humans, has provided a startling portrait of the great whales bounding and playing with the very storm that sends mankind hurtling for cover. He concluded that these storms—exhilarating the whales into mighty airborne leaps above the ocean—were welcomed by the whales as a time to play and strut.

Howling winds, decided Payne, were "a jovial playmate, a source of boisterous entertainment, to a whale."

PAYNE SUMS UP HIS FINDINGS

Roger Payne has the power to separate what is old and known from what is new and unknown. He does this with a mathematical briskness disturbing to the true rhapsodic, but you don't have to be in continuous rhapsody to make some intriguing observations. A knowledge of the higher math helps.

Explaining for the elite of the hydrophonics crowd what can be expected of sound spreading in the ocean, a formal dissertation from Payne would go on like this: "At 20 Hz_o is approximately 0/0003 dB/1,000 yards, a remarkably low value. It means that a transmission distance of 10^7, or approximately 5,600 miles, is required to reduce by 3 dB (i.e., to half the power) the sound energy lost to attenuation!" The amateur student of sound patterns in the ocean might judge, from that final exclamation point, that Payne was onto something rousing indeed. But was it worth it to find out what? Perhaps talking with the family cat would be more suited to one's own personal talents, after all.

In 1979, these were Payne's conclusions about whale songs after more than two decades of field studies: He found that the humpbacks sing during only half the year. Since human speech is not semiannual, that finding works against the proposition that the communication patterns of whales and humans have much in common. He believes he has determined that the singers are males, not females. This has heightened the many comparisons made between whales and songbirds, but it seems to take us away from whalespeak rather than toward it. Although we know that whales are different from us, universally different, to postulate one sex as mute and the other as magnificent explainer-talkers would boggle anyone's imagination.

Breaking the song of the Great Winged Angel into its essential structure, Payne found hundreds of units in each song. The units grouped themselves into phrases, and the phrases became repeating themes. The themes would come one after another, *but*—and this was a drastic *but* to those looking for evidence of conversation—the sequence of themes was always the same. Even though the songs evolve from one year to the next, Payne finds, even in his latest studies, that the whale sings, rests, and repeats. This is not the nature of what we ordinarily understand as conversation.

"This is an inviolate rule of form," Roger Payne said of the cyclical system showing on his charts, "but the number of themes to a song is arbitrary, for some themes are sometimes omitted."

What did it mean, then, and how far had we come—if at all—toward whalespeak?

Whales have a refrain? But the wind has a refrain, the wind even has variety—but the wind does not talk. Looking for some way to square the notion of whalespeak with the repetitive patterns that Payne had noted, those who were determined to find a way to converse with whales could take comfort from the fact that each new season brought new refrains.

Payne had noticed this in the early 1960s, even before he began his own advanced investigations. In Bermuda, another whale investigator, Frank Watlington, had made tapes of the humpback's trilling song style from 1961 through 1964. Later, when Payne and a colleague, Scott McVay of Princeton, listened to the Watlington tapes, they realized that the whales of 1964 were singing a different song than the whales of 1961. They pondered the idea of whale dialects—maybe each tribe of whales has one song and endlessly repeats, and if you hear a different song you're hearing different whales.

Payne went to sea to make some new recordings, and in time the whale was more extensively recorded in the Payne collection than Mick Jagger is by his recording company. Collecting all known fragments of whale songs as well as recording new ones, Payne filled in his collection until it eventually reflected whale songs for every year from 1957 to the present. And recordings are still being made. He credits his wife, Katherine, an artist, with the "fascinating discovery" that the whales in any given area "sing the same song as each other and all keep up to date with the current version of the song."

Now that was a bit better from the standpoint of the claque for Whalespeak Now! Whales with a "hit of the year" were close enough to a human model for some to feel that we were getting somewhere. I think I thought more of the proposition that whale songs resemble gospel singing, where there is much chant, much echoing and re-echoing. This kind of communication reaches remarkable peaks of emotion. Gospel singing can be an effort to sing your way into communication with the Higher Spirit conceived above you.

To a group at the University of Iowa, Roger Payne said:

> The question of why whales sing may in fact have no answer. It is possible the whales don't know themselves. After all, why do people sing? Could you answer someone who demanded to know why you sing? Speech is just one way

to communicate. Another way is through music. People use all art forms including music to communicate feelings they don't fully understand. . . .

If we propose to classify whales in the ranks of human musicians, how vehemently that would be opposed—even though there currently exist several works by serious human composers scored for whale and orchestra and performed, in one case, by every major American symphony orchestra. Whether or not we subscribe to a whale's song as music, it was created by a form of life that composes and does so within a set of laws of form as complex and strict as our laws for composing sonnets—a form of life that has filled the vaults of the oceans with this music for millions of years—filled them with untold arias, cantatas, and recitatives that have echoed and faded and died away never to be heard again.

The Great Winged Angel was a solo singer. It also sang in duos, trios, and quartets, creating, as Payne described it, "beautiful choruses of countless interweaving voices."

There has probably never been anything that so begged for absolute translation as Roger Payne's charts of whale songs compiled in Hawaii, Patagonia, and elsewhere. If you imagine yourself as one of the great code breakers of history, try those Mayan hieroglyphics (they still aren't perfectly solved, at that) or have a try at whale song and see if you can discover what others, including Roger Payne, have not yet found.

Very likely the greatest, and certainly the most thorough, of the whale musicologists, Roger Payne has provided us more to work with in respect to the whale's communication system than we have from any number of earlier and more elusive civilizations of man.

WHALE SONG GOES TO WASHINGTON

Payne did not confuse whale songs—as others sometimes have—with whalespeak, a quite different matter. But although he and other whale naturalists hadn't succeeded in proving the whale to be a free-style conversationalist, Payne had gathered enough clues about the whale's evolving song style to indicate that it was not to be passed off as static and unchanging. His findings were significant enough to come into play at a crucial moment for whales and porpoises.

In the flaming atmosphere surrounding the whale and porpoise massacres of the late '60s and '70s, strong arguments were needed to overcome the economic pleas of whaling interests and porpoise-killing tuna fleets, the richest fishing fleets in the world. Porpoises were being killed at the rate of 500,000 per year as a result of tunamen setting their giant new $200,000 nets around porpoise schools to trap the tuna swimming beneath them. The intention was to let the porpoises get away while securing the tuna, but, in so-called "disaster sets," this method backfired. As many as 800 porpoises could die in a single netting by a single boat. All through history, the porpoises' intelligence and sparkle had saved them from such massacres—until our own generation came.

Insofar as the public understood it, it was a great public scandal. If the public could get its mind on the matter, it objected. Whales and dolphins—they were supposed to be intelligent, weren't they? Bright enough to talk?

The porpoise massacres started in 1959 and remained a sailor's secret until the middle 1960s. They came to light because tuna boats would arrive in San Francisco and San Diego with the corpses of porpoises still aboard. They hadn't had time to push them all overboard. The corpses identified the crime. Anger raced across the country but human voices could not, by themselves, halt this rape.

As for the whales, many of their species were—and are—in real danger of disappearing. The great gray whales, once thought to be extinct, are barely surviving; a few managed to escape the harpoons, and now they are on the upgrade again. The largest whales, the great blues, remain the most endangered, along with the Alaskan bowheads. The humpback, the great singers Payne recorded, were seriously endangered, their numbers dropping within the twentieth century from 111,000 to 5,000. They now have endangered species status, worldwide. The right whales, once the principal targets of the whale fleets, have been protected since 1935 and are making the narrowest of comebacks. The fin whales are down by three-quarters, to between 80,000 and 90,000. Many of this species were destroyed in the 1960s when the whaling fleets reacted to prohibitions against killing the blues, the biggest of all, by turning to the next biggest, the fins. Of the eighty species of great and smaller whales, most are only narrowly surviving, and some—like the pygmy right whales—can barely be found.

By the 1960s, then, the killing of whales and porpoises had reached an all-time peak, and in 1970 Senate hearings were held on the Marine Mammal Protection Act in an effort to get the United States government to ban these widespread massacres. One of those most outraged by the slaughter was Senator Hubert Humphrey. During an impassioned speech on the subject, Senator Humphrey quoted the Russian delphinologist Yablakov as follows: "Dolphin societies are extraordinarily complex, and up to ten generations coexist at one time. If that were the case with man, Leonardo da Vinci, Faraday, and Einstein would still be alive. . . . Could not the dolphin's brain contain an amount of information comparable in volume to the thousands of tons of books in our libraries?"

In Hawaii, a young marine geologist named James Hudnall conceived an argument that I sometimes think of as Hudnall's Theorem. He suggested that the only sure way to prevent humans from totally wiping out another species is to convince them the creature talks. Hudnall's Theorem began to have a concrete effect when whale song came to Washington.

Human voices alone had not been able to halt the devastation of whale and porpoise. It took something else to give the cause the kind of emotional fervor a politician could hang his hat on. Hudnall's theory called for it, Payne's exhaustive experimental studies supplied it: an intriguing demonstration of the communicatory powers of the great whales. And Roger Payne proved to have a great ability to persuade politicians to look into their consciences and seek to bring the long travail of the whales to an end. There was one moment, during the marine mammal hearings of 1970, when Payne looked at the distinguished members of the Senate Subcommittee on Oceans and Atmosphere and told them, as though to buck up their spirits, "Let us be brave . . ."

It *took* some bravery to fight the tuna and whaling interests. They had clout.

Payne and those fighting with him triumphed, and whale song played a part. Payne's charts went into the record of the hearings. The evidence thus afforded of the creative intelligence of whales helped to bolster the pleas for saving these animals. Should we have needed to prove extraordinary intelligence in order to keep the whales alive and give the porpoises protection they lacked? Perhaps not, but it helped. The idea that there might be a talking-bee out there in the sea provided a powerful motivation not to let the whales vanish completely.

In retrospect, it seems even clearer than it was at the time that the *argument from intelligence* had been the saving of the whales. *Whales talk! Either that, or they have a hell of a national anthem!* It was a view that politicians were ready to buy once they believed the mass of the people had accepted it. Whale songs and all the accompanying studies have already played a historic role. It's not the first time that the tide of battle has been swayed by a song (the "Marseillaise" always seemed to help the French), but it's the first time that the song which helped the marchers march came from a whale.

So whale song and porpoisespeak, with no translation available, had achieved a sort of third-dimensional reality right there in Washington. The whaling industry, in America, was at an end. The tuna fleet fought on but lost. Waste no tears on the tunamen, however. They rescued themselves later, when, with the law already passed, they began to lobby for special permits so that the porpoise massacre could go on. After ten years, though, the annual "accidental kill" of porpoises is down from 500,000 a year to under 20,000. A rising avant-garde in animal communication theory can claim at least a partial role in achieving this reduction. Until there was a large-scale public belief not only in the dangers of extinction but in the prospects for intelligent communication with these nonhuman species, the troops had never been strong enough to carry the day in Washington.

TOWARD A ROSETTA STONE IN CETACEA

Despite the thousands of words that have been written on the prospects for "breaking the code" of the great whales, this remains a hope and not a fact. The interspecies communicator can feel, as many do, that a human reacting esthetically to the songs of whales is receiving more than any ordinary conversation would provide, but that isn't the same as having a back-and-forth discussion.

We do have some developing evidence, however, that whale-to-whale communication is itself sufficiently intense and diverse to suggest something akin to language. Dr. Lilly's seemingly bizarre ideas that whales can exchange historical tales and social notes should be held as an open question.

Ken Norris told Michael Parfit that it was purely delusionary to think there is evidence that porpoises or whales have the ability to "carry historical and ethical information from generation to generation. There isn't anything that even *hints* at language," Norris added. "What I see instead is a great rat's nest of sound which nobody—*nobody*—knows about."

Although Ken Norris could prevail in the end—the issue is not closed—at least one case has built up now that *does* seem to provide hints of language and may be quite a bit more than just a rat's nest of sound.

The capture of the killer whale Namu (scientists use the name *orca* for the species) is certainly one of the best-known animal tales of our times. It has been little noted that an investigator named Thomas Poulter moved in on this unusual capture, hoping to take some recordings of orca sounds, and stumbled into a more exciting and revealing situation than he had expected.

When the Namu capture occurred in 1965, Poulter was senior scientific adviser at the Stanford Research Institute and director of its Biological Sonar Laboratory. After taking a doctorate at the University of Chicago and advanced science degrees at Iowa Wesleyan, he'd had an incredible career, including a role as second in command of the second Byrd Antarctic Expedition from 1933 to 1935. He was sixty-eight years old at the time of the Namu incidents and was in the process of compiling a master study of the communication patterns of all the marine mammals. In the report contributed to Thomas Sebeok's massive *Animal Communication* anthology, Poulter had noted that there had "seldom, if ever, been an opportunity to study the vocalization of the killer whale in a systematic manner." He would soon be 60,000 feet deep in recording tapes, and detecting an amazing story from their unreeling.

The saga of Namu started on a blustery June night off the British Columbia coast. Two fishermen caught in the storm snagged their net on a reef during the blow. They abandoned their net to get under cover and didn't come back until the next day. Their own misadventure led to a greater misadventure for an adult orca and a baby killer whale as well. The larger whale, later called Namu, could work its way in and out of the net but the baby wouldn't follow. True to some sort of ethic, the larger whale would not abandon the baby. The fishermen realized they had something and looked for a way to make money out of this odd affair. It happened to fit in neatly with the long-term quest of Edward ("Ted") Griffin, owner and director at the time of the Seattle Public Aquarium. Griffin had been trying for years to capture a whale, calling it his "ruling passion." How much of the passion came from curiosity and how much from a showman's realization that he would have one of the stellar attractions of the world is a question only Griffin himself can answer. It might be guessed the pull was strong from both directions.

Griffin had been out with "boats, helicopters and tranquilizer guns," trying to take a whale alive in Puget Sound, but it hadn't worked out. Now he was the beneficiary of a fluke. He put together $69,000 in a hurry and created "the Tinkertoy flotilla," named for its flimsiness, which was to bring the captive Namu down the coast to Seattle. At the beginning, Namu was in a completely vulnerable system, awaiting transfer to a pen they built. When a force of forty killer whales turned up, presumably responding to distress cries of the captive, the Tinkertoy works was not much of a bet to survive an attack.

A force of killer whales coming in a rush can be the most chilling sight in the sea. They are a more formidable force than a frenzy of sharks. It's quite true that the orcas do not attack humans, but they attack the great baleen whales, who are far larger than they are, by biting at their mouths and seizing parts of their tongues as snacks. There is nothing genteel about an orca attack, and the force bearing down on Griffin's small party must have been terrifying when they first hove into view.

A really sharp commando team of orcas could have gnawed the Tinkertoy works to a shambles, freed Namu, and scattered like Israelis from a raid. But they didn't. Whatever their degree of braininess, the orcas couldn't figure out the strengths and weaknesses of a human engineering system. The protections for Namu were duck soup to penetrate, but the orcas never guessed it and Namu never guessed it either. The whales charged the nets—and stopped. Charged again—and stopped. Why?

Griffin concluded that Namu could have been imprisoned, with no hope of freedom, in "a sack of tissue paper." The whales, Griffin concluded, depended on their sonar system, and the system, while it detected the barrier, didn't tell them how vulnerable it was.

Namu was shifted to a stronger, still makeshift pen for the long journey down the coast to Seattle. The distress calls from the captive kept a siege of orcas near the flotilla. This attack force, or fanfare of whales (you could look at it either way), was beyond the wildest dreams of the most hyper aquarium publicity man. A cow and calves joined the trailing orcas, and reporters rhapsodized over the "family" that couldn't bear to see Daddy shanghaied.

Once Namu was safe in Seattle and wintering in a cove on Rich Passage, contained by a submarine net, Thomas Poulter began to amass night and day recordings. Sleuthing his way through the sounds on the tapes, he couldn't have predicted, at the outset, exactly what this was about to demonstrate in terms of whale talk. When the female killer whale Shamu was added, it provided one of the elements needed to see if there was significant communication. They would analyze what the whales said to each other. But a surprise was in store. Poulter soon realized they had recorded not only the captive whales but noncaptive orcas as well. He concluded that the exchanges were taking place between captives and noncaptives at distances up to seven miles.

Poulter deduced that something extraordinarily complex was taking place in the sounds passing back and forth, over a range of miles, between whales who seemed to have a great desire to stay in touch. It was the complexity and variation of the sounds that led him to describe them in a different way than he had the instinctual sounds of all sorts of marine mammals he had previously catalogued, from the bearded seal and white-capped sea lion to the narwhal. While the orca exchanges might have been interpreted as similar to the baying of one coyote to another, Poulter was excited because he felt something deeper was implied in the tapes. Describing the individual signals as lasting from half a second to five seconds each, he provided figures on their two-octave range and then suggested, "The male killer whale therefore has an extremely complex signal framework which can be recognized against almost any background noise and which he can accent, abbreviate, punctuate, syllabify, hyphenate, prefix, and give numerous endings and inflections without affecting its ease of recognition. . . . I suspect the different signals do make sense to other killer whales. . . . It is believed significant that many more modifications of Namu's signals occur when he is exchanging vocalization with a female killer whale than with another male."

Poulter concluded that there was a statistically valid case for "animal language" based on the Namu-Shamu tapes. The deliberation with which he insists on that mass of punctuating, syllabifying, hyphenating, and prefixing indicates how strongly he is asserting *that really, now, this is a very complicated sort of communication indeed.*

Why, then, shouldn't the Namu-Shamu tapes themselves become the Rosetta Stone? Let the computers work on them, coordinating them with vast transcriptions of human talk and additional sounds and signs from killer whales, until equivalencies are worked out and we have translated the language of the orca. As neat as this may sound to the interspeciesist, Poulter didn't think it would happen. Not only were the killer-whale signs complex, but they were *so* complex that he couldn't imagine a whale vocabulary being sorted out for comparison with a human vocabulary and a whale syntax being lined up with the human syntax.

"If we had the nearest equivalent that exists in the English language for each of those signals," Poulter said, "I suspect that we would still be orders of magnitude away from being able to make any combination of them that would make sense either to us or to the killer whale."

After a long, quiet, closely reasoned essay on his findings, Poulter ends with what amounts to a cry of triumph. "Yes!" he suddenly affirms. "We believe that marine mammals talk and that what they talk about makes sense to other marine mammals of the same species."

If we are right in assuming great complexity—not simple signaling—in the Namu-Shamu tapes, the position is somewhat the same as for the Mayan codexes (hieroglyphic scrolls). When the Spanish burned the Mayan scrolls in a great pyre intended to abolish pagan thinking, they were astonished to see the Mayans weep as the flames destroyed a library that had been built up over many centuries. What the Mayans had to say in those scrolls must be much closer to our own tongue than the expressions of whales, and yet the best of human intellects working with a few surviving codexes fought fruitlessly for most of five centuries to effect a translation. Only in the last few years has it been claimed that the Mayan scrolls are "giving up their secrets." Many a Mayan scholar roundly condemned the Spanish because they had not salvaged a translation at the very beginning. There were, after all, still Mayans who could have read them when the conquests began. Later, not a soul was left on earth who understood.

Since killer whales were not on the whale fleet's hit list, a comfortable number still exist in the world. It is still possible that if we are unable to make a translation of the Poulter tapes ourselves, we may come into an age when we have conveyed some version of our own language to an orca. And the whale, in turn, might go over the Namu-Shamu tapes and explain their content to us.

The case seems not quite so desperate as somehow losing out on that Last Mayan.

WHALE SOUNDS INTO SPACE

The Australian Inquiry into Whales and Whaling, a privately funded documentation aimed at putting a deep perspective on whaling issues, was influential in altering the course of whale protections afforded through the International Whaling Commission. The inquiry came to play a major role in discouraging further whaling. It placed considerable stress on the possibility of "high intelligence potential" in whales as a reason for sparing them. At the same time, a leading Russian authority, Dr. A. A. Berzin, argued that "the sperm whale brain structure is such that this can be said to be a thinking animal capable of displaying high intellectual abilities."

How high? At least one gamble has been taken on the notion that the whales' mode of speech may be quite as impressive in outer space as our own.

Voyager I and II, in their billion-year space journey, are carrying whale sounds as well as human sounds. Should contact be made with inhabitants of other civilizations, other universes, they will have these samples of earth sound to judge us by.

The interspeciesist would say: *We have absolutely no reason to think that living beings in the outer reaches of space are more certain to recognize the ideas of the humans than the ideas of the whales. It's always possible that nonearthlings will feel closer to the whales than they do to us.*

MRS. BENCHLEY EXPLORES A PATH THAT LEADS TOWARD APE TALK

From the early days when she greeted exploring friends like Osa and Martin Johnson (holding lion cubs) through a long series of experiences with remarkably intelligent and communicative gorillas, San Diego Zoo's long-time director Belle Benchley looked for clues to what the animals were saying. She found gorillas the most expressive of all the exotic creatures who came under her charge. Belle's big armload below consists of the zoo's first baby lowland gorillas, Bouba, Albert, and Bata. *(Photos by the Zoological Society of San Diego.)*

DOLLY LEARNS TO BE A MOTHER

An unwilling mother at first, Dolly of the San Diego Zoo was given instructions in motherhood by young primatologist Steve Joines. Using gestures, words, and a gorilla doll-baby, the human teacher gave Dolly the lessons in mothering she would have obtained in the wilds from older female gorillas. After a series of experiences with Dolly and other gorillas, the professionally trained Joines concluded "Gorillas understand English." The above photo shows Dolly with her baby, Binti. *(Photo by Zoological Society of San Diego.)*

THE MELONHEAD: A RELUCTANT COMMUNICATOR

Sea Life Park's Ingrid Kang, an adroit trainer of those small whales usually called porpoises or dolphins, needed all her skill to make contact with the small melonhead whale, Lahaole, shown here. Friendly by the time this photo was taken, the melonhead—one of a pair—stayed under water long after its capture. How the trainer learned to communicate with the melonheads and bring them to the surface is one of the unusual stories to emerge from recent annals of the oceanarium trainers. See chapter two for the full details. *(Photo by Nicki Clancey, courtesy Sea Life Park, Makapuu Oceanic Center, Waimanato, Hawaii.)*

DR. LILLY TRACKS AN EXTRA-TERRESTRIAL THINKING FORCE – THE DOLPHINS

The most brilliant expounder of the case for interspecies communication, Dr. John Cunningham Lilly organized the experiment in which Margaret Howe (shown in pool preparing for

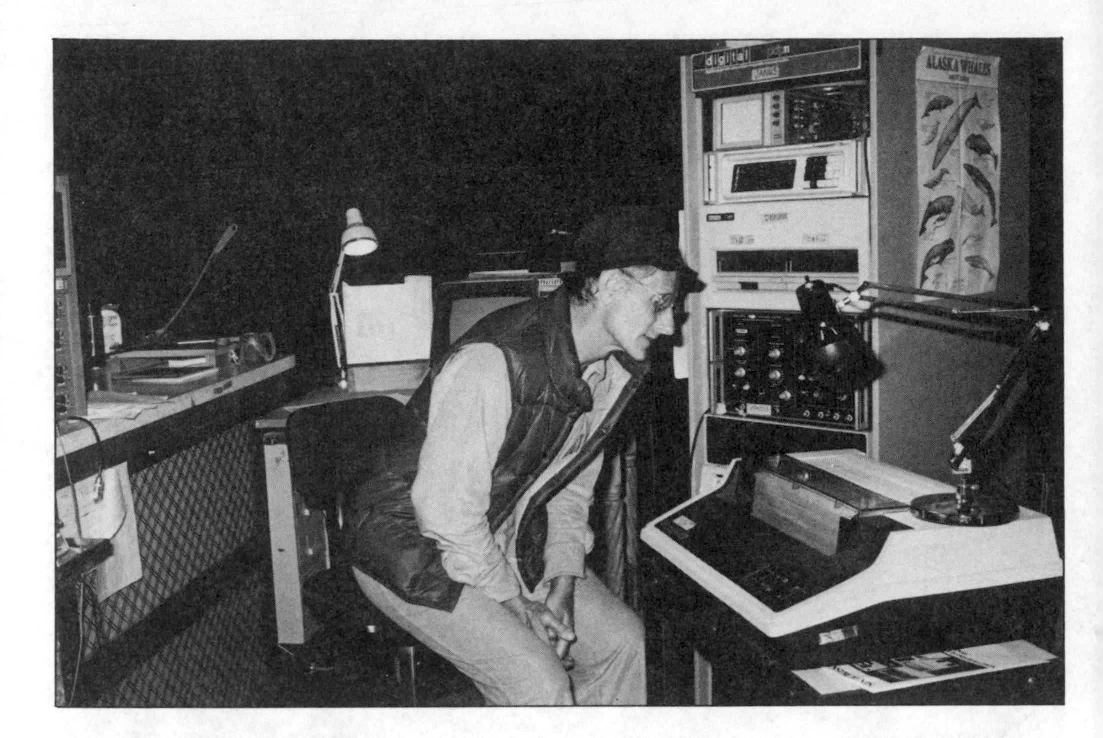

a language lesson) lived night and day with a dolphin to increase the sense of closeness between species. Lilly and wife Toni confer with a dolphin who has come out of the pool at their urging. By the late 1970s, Lilly was using the Janus computer in an effort to translate what he refers to as Delphinese. *(Margaret Howe photos courtesy the Communication Research Institute. Dr. Lilly photos courtesy Human Software, Inc.)*

THE REACH FOR
THE DOLPHIN'S UNIVERSE

Margaret Howe, pictured above with her confederate Peter, came as close as anyone to actually surviving in the dolphin's own element, by living for several weeks in a Lilly-planned wet-room. But the human can take up a water existence only on a very partial basis, and the dolphins can come above the water, into the human's world, only for moments. Investigators of communication technique are still trying to overcome this basic difficulty. While dolphin and whale may have a greater communication potential than the primates, they are harder to reach because they exist in what is essentially another universe.

THE WHALE COMMUNERS: A GLOBAL SEARCH

In skindiving gear or in small boats, bands of hopeful whale communers advanced into whale country, trying to make contact with these large and still mysterious creatures. So rare that it was rumored to be almost extinct when the whaling mounted to its historic peak in the 1960s, the gray whale (pictured here) is making a reappearance now. Sometimes it gestures the communers away—but if the interlopers seem friendly enough, the gray will also come up and allow itself to be fondled. *(Photos by Ronn Storro-Patterson.)*

PIPER TO THE ANIMALS

Jim Nollman became a serious exponent of the view that animal consciousness can often be reached via music. In the scene pictured at left, Nollman plays a waterphone for wild killer whales who accept his presence. In British Columbia Nollman recorded a series of exchanges in which whales seem to be trying to teach him the proper order for a string of notes. He regards this success as a first step toward active communication between humans and whales. Nollman also used his agility with ocean acoustical effects to try and save the dolphins who are murdered as "gangsters" by the Iki islanders of Japan. Nollman's appeals to the Japanese people on behalf of the dolphins were aired on television, below, with Japanese titles translating his words. *(Photos courtesy of Jim Nollman and Interspecies Communication.)*

14

Jim Nollman: The Leader of the Pack

YOU CAN MAKE THE ANALOGY," JIM NOLLMAN SAID, "that flying and talking are the same thing. Or the same degree of things."

What in the world could the man mean? If there are two things on earth that are not remotely the same in purpose, in effect, in style—in *anything*—surely they would be talking and flying. You could sort through ten million people and find it difficult to discover even one who would tell you that flying and talking are . . . well, look at it this way, sort of the same thing.

But then that's what Jim Nollman is: one in ten million. It has made him the leader of the pack in that small but contagious body of persons who try to teach us that communication is a concept broader and stranger than we, with our patterned minds, had ever imagined. Because Jim is so unpatterned in every way, his mind slips upward to the idea that in the manner of the bird's flying is the sparkle of the thought. While others tried to crack the code of birdsong, Jim followed this idea of bird flight and what it might all be about.

He doesn't claim to have unraveled it, and he is not a systemizer, recording tens of thousands of bird swoops to see if he can some day translate one of those swoops into "Hi, my name is Sophie."

Let others argue the proofs; Jim himself, like a composer or an artist, tries to release himself into the spirit. Once you grasp his idea of flight as a conversation with the universe and with its other inhabitants, this quite unthinkable idea becomes very thinkable. Soon it is difficult *not* to believe it. You go back to watching birds in the sky with a fresh idea of what their flight is about. Sometimes, of course, it is chase-and-peck. Sometimes it's utilitarian travel, no more meaningful than a workman's ride on a bus. Sometimes it's stuka-diving—pounce on your prey and have dinner. But there is another kind of flying—stunt flyers and barnstormers must have sensed it—that speaks of exultation, drifting thought, the comedy and righteousness of one small being against the cosmos. In its endless rush and dashing, in its rising arc of flight or swift sprints and skims, the bird is expressing itself as fully as a ballet dancer or a cheerleader.

"I think seagulls communicate with the wind and with each other," Jim has said in an interview with an off-trail magazine called *Magical Blend*. "As a musician, I don't think language is a higher form of intelligence. . . . If you were a person and you wanted to establish a relationship with someone from Alpha Centauri, you would probably think of language because that's the way you communicate, but if you were a seagull you might fly. Seagulls might think that we're totally off the wall. How stupid can we be, not to be able to fly?"

In the half dozen years since the determination to break through the animals' communication codes in a personalized way became a very generalized obsession, a growing number of the explorers in this field have become aware that an ambling musician who has fewer formal credentials than any of them has actually been getting somewhere. With his strange twists of thought, his openness of mind, and his dexterity with any instrument that will croon, croak, gark, or wheezle, Jim seems now and then to cross the barrier and actually pick the animal's brain (or let it pick his).

DOLPHIN TALK: THE TRANCE APPROACH

Certain people are a rumble on the horizon long before you meet them and before they have become famous to the world at large, and Jim Nollman is one of them. Wherever I went among those who believe that conversation with animals is passing from theory to fact, someone would say, "You know, there's a fellow named Jim Nollman, who . . ."

Jim Nollman has proved to be the closest thing to a Pied Piper that exists in the world. He often seems broke, without visible means of support, a circumstance that befits a Pied Piper. It seems not to worry him as it would other people. Every few minutes, his mind strays to the animals he hasn't spoken with yet. Usually, when I hear from him, he has just popped in from continents away. He is the director of a three-year-old group called Interspecies Communication, which issues from Bolinas its own minuscule-circulation publication with Jim as its star and principal communicator. In California, where cults spring up like toadstools after every rain, one more should make no difference, but it has.

In describing the methods of musical communication which he explores—in far different fashion than Roger Payne—Jim distinguishes between the "mirror approach" of his friend Paul Winter, a saxophonist with perfect pitch and an almost preternatural knack for matching an animal's exact intonation, and Jim's own "trance approach." The mirror approach is a way of arresting an animal's attention and working toward an exchange of meaningful notes, but it has an easily discernible drawback. If you say "Me Tarzan" and I repeat "Me Tarzan," we are not necessarily going to get anywhere. Jim's trance approach is another way of saying "go with the flow"—he simply insinuates himself into the animal's consciousness, trying one sound and then another, sometimes one instrument and then another, sometimes mimicking but also departing from the mimicry, letting one reaction take him to another, intriguing the animal with musical phrases that fit and yet surprise, experimenting, experimenting, shifting and shifting until—until, at times, an intelligible reaction results.

He has tried this with greater agility and inventiveness—and with greater success—than anyone alive.

It is clearly a task that only the most patient man in the world would undertake, and Jim, I suppose, could win that honor. A very odd mix of patience and unpredictability has led him on a zigzag course toward fulfilling Dr. Lilly's prophecy. Lilly *wants* to go to sea to test out his theories, but it proves easier to work with captive porpoises under controlled conditions. Jim wants to go to sea and actually gets there. Although his real breakthrough is with orcas (and his greatest feeling of "oneness with the pack" is with wolves), he has also found deft ways to beckon and then infiltrate the tribes of wild porpoises that Lilly believes are the most likely subjects for back-and-forth communication.

Cousteau has said that despite their friendliness in captivity, wild dolphins scatter from skindivers and cannot be approached.

Yet here is Jim on one of his many forays: He has strange encumbrances when, a lone swimmer, he pushes out to sea. A buoyant bobble of brass and steel moves in front of him, and he grasps a mallet in his teeth. When fins appear in the distance, Jim works his machine to produce vibrations that shoot through his body even as they shoot through the ocean. This is not like tuba playing; this spherical instrument, with prongs rising from its center, sets everything around it, including Jim, vibrating. "The whole process tickles," he says.

This instrument has a deceptively appropriate name—the waterphone—but this doesn't stand for what you might think. It was invented by Richard Waters, who, says Jim, decided to name it after himself.

Does it have the desired effect on the dolphin? At first, there is no sign that it does. He has gone 200 yards offshore and those fins are as distant and unconcerned as before. Jim toils alone, marooned with his obsession, trying one set of vibrations, then another. He is looking for the sound and rhythm that will tell the spinner dolphins that he is a speaking and sentient creature. The dolphins pay no attention. Time passes and Jim's mind drifts. Someone else would react to this lack of interest by going back to the seaquariums and dealing with those dolphin who, in their bored condition, should be glad for any intrusion. But Jim is stubborn—one of the great waiter-outers.

He has begun to conjure up memories of movies about predatory sharks, and it's the last thing in the world he wants to think about. Jim is mad at his own brain for its unruliness—for being, as he claims, "media-cluttered." It's true that he is in no condition to be thinking about sharks. Far from shore now and juggling an instrument that looks like it's out of the Space Age—and is—there's no way he could handle a shark attack. If one should come zooming in, about all he could do is say "Please don't"—or see if there's some note to produce from this vibrating globe that means "bug off" in shark language.

While all these untidy thoughts race through his mind, Jim continues to experiment with sound and rhythm, watching for any change in the attitude of the dolphins, who still seem far away and not inclined to bridge the distance. As many as twenty fins cut through the water, none paying him any heed. Then the mallet, hitting the center tube, produces an odd rising and falling pitch that—like some message from a distant planet—has, in a moment, caused the dolphins to regroup. They are suddenly afire with his presence and, as he repeats the note and varies it, they form into tight circles and begin a spiraling dance. Now Jim becomes the vortex instead of a small plea in the wilderness. Whirling in tighter, madder circles as they respond with an inner music to the tones from his spacy gadget, the dolphins remind him of Israeli folk dancers.

If you've ever seen the dark Israeli girls come out of the shadows at a party to hurl themselves without restraint into some fiery rendition of "Hava Nagila"—the message of the song is "Run to the mountains and rejoice!"—you have some faint notion of what Jim has suddenly sparked from a quiet sea. The mood seems crystal-bright and joyous, and the dolphins are a triumphant troupe of wild dancers.

But then the mood changes again as the dolphins break from the circle, one by one, coming closer and closer to him, inspecting the stranger even as they respond to him, poking their heads above the surface to take a long inspective look.

The one particular strain of sound that Jim has discovered carries a different message for the dolphin from the very similar sounds he had been playing before. Hard to parse the mystery of that—what message does one sound connote that another doesn't? And yet it is true for humans, too; there is music that sets us jumping and other melodies that lull us toward sleep.

Jim Nollman has found the key to bring the dolphins all around him, making him feel safe and greeted. As the bolder, inquiring ones keep sticking their heads above the water to fix him firmly in their sights, his mind registers a different thought than might have been expected. As he wrote later:

I am mostly struck by the immense power of their breathing. All the time I continue to play these long sliding notes on the waterphone. Once more I stick my head into the water. The sphere sounds clear and sweet: now like an oriental gong, now like a church organ. I listen for twenty seconds, come up for a deep breath of air, stick my head full below the surface again, up and down, over and over. This process of playing and listening seems so clumsy. I feel inadequately built to the task of communing with these creatures of the sea. I still can hear nothing but the sound of the waterphone. Perhaps my ears are just not built to hear the high-pitched sounds the dolphins may be making.

But if I feel inadequate, the dolphins don't seem to notice it at all. All of us are quite together now, human and dolphins swimming on the deep blue sea. One swims directly underneath my feet. And now, so suddenly as to make my blood rush, one of the dolphins jumps six feet clear of the water and just a few feet away. In a moment they are all jumping clear of the water, spinning and somersaulting about. All I can manage to do is watch them with a big foolish grin on my face. And from the shore, so very far away, an audience of humans has gathered to watch this interspecies theater. And now they too are all jumping about, laughing and clapping, carrying on like a group of happy children.

This joy is truly infectious. What more could any musician ask for from a performance?

"AND, BY THE WAY, BE A SAINT"

No one who is aware of Socrates' little butter-dip of a nose, or who remembers de Gaulle stalking haughtily in his garden for years at a time so that the world would one day notice his funk and call him up to command the nation—no one, that is, who is at all sensitive to the many forms and styles of the actual visionaries of this earth—will be very surprised at what I have to tell of Jim Nollman. In a visionary, expect odd packaging.

I wonder if those wolves, dolphins, and whales who have made his personal acquaintance are aware of the extent to which Jim Nollman differs from the rest of mankind.

It's not the voice that confounds you—at least not immediately—when Jim steps into a room. The voice is expectable and pleasing. It has a sort of upturn in it, a pleasant warble. His speech is constant and inquiring, there is a sort of plea in it, always, but it also has the prophet's feel for the convincing note. Jim's voice could be calculated to move you toward trance if he were to choose that. No, it's not in the voice but in his appearance that Jim shocks. When first he hove upon my sight, he looked, as someone said later, as though the seediest San Francisco commune would reject him on grounds he wasn't up to their dress code.

That's a misimpression. The shirt-top that looks like somebody's lost underwear proves, on closer inspection, to be just new-wave rather than ragtag. Maybe Jim is out in front on clothing, as he is in talking to animals, but your opening impression is that he must have had a bad day pulling tugboats through the swamps.

After we had talked a while, I led Jim through a corridor to meet a gentleman who is meticulous about clothes, his own and everyone else's. His face didn't give a flicker, but I am well aware that he can spot a hangnail at thirty paces. In this carpeted office, Jim was as odd as somebody who has stepped out of a Picasso painting, but his murmuring voice with its quick rhythms and special flights was already at work on the awesome fellow before him. If you close your eyes on Woody Allen, say, you will realize that it is the voice that is melting away your impression of his face. Allen's voice is as ingenuous as a puppy's squeal, and so is Jim Nollman's—the compulsive murmur surrounds my friend in the big desk and then takes him over. Soon there was no sign that the two of them were anything but fellow enthusiasts, perfect partners for great ventures, soulmates.

The Pied Piper's voice is there so fully that, even as he can intrigue a porpoise, Jim can throw out the notes which pique serious and somber men into the notion that we do not have to live in a workaday world, that there are visions to contemplate and new ideas to try. Something else, too, is always present in Jim's continuous soothing and surprising chatter. He shines with goodness. As he talks, I'm reminded of that casual instruction Baltasar Gracian y Morales offered us back in 1647: "And, by the way," he said, "be a saint."

So the animal communicator I thought would come like a superman or those Viking sea divers I had been meeting resembled instead a Bowery messiah. He was soon confiding, in that intimate patter which makes people dance around him even as the porpoises do, that his career as musician to the animals had started on a turkey farm.

Living among Mexicans who often keep turkeys as pets and trying to master on a flute what he calls "Central American scales," he would hit certain notes that caused a flutter in the turkey in the next yard. Jim noticed what was happening and began to influence the turkey with his flute. If he hit the right note, the turkey would come flying over the fence to do a dance for him. It might have remained just some offshoot of Jim's quirk for delving into obscure musical patterns, but he happened to mention the turkey's responsive behavior in a letter to a talk-radio host in California named Charles Amirkhanian. By the time Nollman was back in California, Amirkhanian had conceived an idea for a Thanksgiving special—Jim would go to a big-time turkey farm and set the whole flock agobble with a few notes from his flute.

There wasn't much reason to think it would work, but it did. And Jim, far more startled than the audience who happened to hear this first Gobble Festival, began to do some serious thinking.

"I couldn't get over it," he said. "If you spoke to them in the right way with a flute, the turkeys simply dropped everything else to sing and dance in response to what you were saying. Now that's odd. I guess the people who heard it took it as a Thanksgiving novelty and forgot it, but there was a ticking in my ear that said, 'This means something, follow it up.' I kept thinking over what had happened. When I played a note, 300 turkeys would respond together. It was a little like a basketball game. Somebody sinks a basket, the whole crowd responds."

A writer as well as a musician, he would run across bizarre hints that remote peoples of the earth—living closer to nature than city folk, who don't even have time for pigeons—have at times mastered certain kinds of interaction with animals; and then lost perspective on it as wars rage, liners and railroads intrude. The old talesmen claimed that Greek fishermen had been befriended by dolphins. The dolphins would bring them fish and the fishermen would give them rewards, make friends—and call the dolphins when they wanted them. The human race, it seemed, had been talking to dolphins in a perfectly practical way ages ago but had lost the knack. It is a knack that can be easily picked up; Jim started the process anew and so did other inventive folk in different parts of the world. But he was thinking about a fuller kind of communication than the go-and-fetch systems that appeal to the practical-minded.

A TRY FOR THE KILLER WHALE

If Thomas Poulter was right, there's more conversation out there in the ocean among killer whales than there is among other animals. But can a human beam in on the conversation? Jim Nollman had a try, using his own methods, and the result seems unique—so far at least—in the annals of interspecies communication.

He had been near whales in the ocean often enough to believe that they were fully approachable.

Jim thinks nothing of it when he has the waterphone buoyantly before him and killer whales are coming down on him as though he were some stray chunk of meat. They aren't sharks, and he expects them to dive for deeper water just before reaching him. So far, they always have. A thirty-foot orca can be as voracious as the great white shark, and it sometimes is when it's involved in gang-style attacks on the great baleen whales; but for some reason it doesn't go after humans in that way. Whether for moral or dietary reasons—or because playmates of a different species are not that easily found—they have never given Jim a shadow of fear.

Off the northern part of Vancouver Island, in 1981, with a waterphone and an electric guitar that had an underwater speaker, Jim finally made the contact that seems like the first positive, provable, fully documented instance of human and whale moving into the kind of verbal communication that means pieces of language are being transferred between them.

Sounding the guitar and the waterphone and listening to the orcas through underwater microphones, Jim started an open-ended experiment. Anyone can hear his results on the tapes he made of the whole thing. He offered me this description of how the recording was obtained: "It was different from all the other experiences and different from bringing the dolphin into a dance, even though you can be pretty certain that there are ways to repeat that effect with other dolphin in other places," said Jim. "I get in the position of having this song—a song I invent as I go along—and the animal responds and you find, by trial and error, what songs he responds to. But the orcas reacted in an altogether different fashion.

"They taught *me* a song, and when I turned out to be a slow learner, from their standpoint, they worked with me until I knew what I was doing and could duplicate it.

"The underwater microphones proved what was happening. We couldn't see the orcas but we could hear them. And they started coming for a rehearsal every day at the same time—about 10:30 at night. I'd begin playing and they'd begin singing. Then we were building along together, the orcas giving me a chance to repeat to show I was getting it. It went to two notes, three notes, finally to four notes. But it was the idea that they had to go back and do something in half-time—show me the steps over again, slower, if I made a mistake—that told me we were in the middle of a lesson and the lesson was for me, the orcas were giving me language. And so I knew they were talking to me. And we have all this on tape; you can prove it to yourself.

"So it's not a matter of whether we can get through to them or they can get through to us. It's started and the way to carry through is just to do it more. Three weeks at a time is the longest I've ever been able to carry on that kind of experiment. It isn't long enough. You start but you don't pursue. Somebody will have to be out there for years at a time, with their sound studio on a boat, and just work with the orca and be very flexible and see what happens. They'll have to lead, they'll have to be led—they'll have to go with it. Let it happen, try to understand it."

It can't, says Jim, be a purely human-oriented experiment. If you want to speak with an Ethiopian in his own tongue, you can't do it by cramming endless amounts of English down his throat. *Orcinus orca,* much as it might want to develop a deeper relationship with these other speaking creatures, does not have human technology, so it will be far more practical for humans to push off into orca language than the other way around. Nor is it likely that the waterphone and the electric guitar are the gadgets that will take us the rest of the way into the language of the whale. They have provided a start, but other ideas and other instruments will appear. Part of performing the experiment successfully is recognizing that orcas desire the contact. This, Jim felt, was proved at Vancouver, and it has been proved often enough before. The orcas readily fall in with human plans, but the humans involved have usually thought small: *throw 'em a fish, teach 'em a trick.*

ON TO IKI

Jim Nollman's idea about the Iki Island dolphin massacre was not to hound the fishermen into giving up the annual bloodbath they were engaged in, but to look for a way to use the new knowledge about communication with marine mammals to lead the dolphins out of danger.

Somebody needed to do something. Events at Iki Island, a fishing center 1,200 miles southwest of Tokyo, had shocked the conscience of the world in 1978 with reports that dolphins were being machine-gunned to prevent them from feeding on the fishermen's catch and that the beaches were red with the blood of 2,000 slaughtered porpoises.

The massacre was real and happened each year, although the machine guns disappeared from the story. When the press services broke the news, the shock waves traveled everywhere, inspiring protests comparable to those directed at Canada for the clubbing of baby harp seals. The islanders, convinced that they were killing "gangsters of the sea," as they called the dolphins, didn't trouble to hide the kills; they were proud and unrepentant even when subjected to international attack.

Trouble between the fishermen and the dolphins had been building for years. The massacre had actually started in 1910, but it started small—annual kills from the earlier part of the century were estimated at ten to thirty dolphins. By the 1970s a fleet of 1,500 boats had concentrated in the Katsumoto Fishery and the fishermen didn't like losing a single fish to the dolphin. Since dolphins had been fish hunters there before the Stone Age, few outside Japan believed the islanders to be anything but heartless murderers. It may not have been as simple as that from the fishermen's standpoint, but the relentless way in which the killings were carried out was well calculated to bring on fresh protests every year.

Jim had two ideas on what might be done. He hoped either to talk the humans out of the massacre or to talk the dolphins away from the fishing areas. Nobody in the Iki Islands conceded that dolphins might have equal rights, prior rights, or any rights; they saw only creatures who snatched their livelihood from under their nose. The cry of "Gangster!" made them want to kill every dolphin in sight. And they did.

Joining with Greenpeace for a 1980 try at stopping or at least holding down the massacre, Jim's Interspecies Communication group pooled ideas for trying to relieve the situation. A fisherman told them that by the peak months of March and April so many dolphins would come that "people could walk from Iki to Tsushima on the dolphins' backs."

Jim believed—and believes now—that an acoustical method for diverting the dolphins could stop the Iki bloodbath. But the government, anxious to show the fishermen that it was standing behind them in spite of international tumult, put an $80 bounty on the head of each dolphin, providing an overwhelming reason to keep the massacre going no matter what solutions might be found to stem the dolphin invasion.

Greenpeace fieldworker Dexter Cate was Jim's companion on the mission to try and intervene. In Tokyo, Jim and Dexter easily made points with the Japanese populace at large, but the Iki fishermen wouldn't budge. For one thing, they weren't the "simple fishermen" pictured in the news accounts. An Iki fishermen is a modern technician with an up-to-date arsenal of equipment—not a workman but a businessman. The fury really derived from the fact that, even with the finest of modern equipment, the dolphins were outwitting them at every turn. When Jim talked with the fishermen through interpreters, wanting to persuade them of the practicality of some of his ideas for deflecting the dolphins, he found them to be "shrewd and technologically articulate businessmen—they use mostly twenty-ton boats that are outfitted with radar, sonar, and many thousands of watts of lighting to illuminate the night ocean." The lights that attract the squid also attract dolphins. Although the fishermen's hooks may get to the squid first, dolphins are masters at filching their prey from the hooks. With the cuttlefish in ever shorter supply, the fishermen were becoming more and more dependent on a fish called the buri—and the dolphin ate them, too. It was maddening. And it was new.

A decade earlier, the fishermen had had swimming "policemen" who kept the dolphins away: killer whales, who eat the buri but also eat dolphins. It took only a few killer whales to make the dolphins depart for other territory. But when commercial fishermen and commercial fishing practices peak, a survival-of-the-fittest contest in the ocean moves with wretched speed. One year, the killer whales were gone—and they stayed gone. Chances are that the Japanese penchant for whaling had simply decimated the last of them around the islands of Japan. Buri were also on the downgrade because Japan's so-called inland sea, where they reproduce, was as fouled by industrial pollution as similar areas in the United States. Pollution from factories is not purely an American phenomenon.

The fishermen didn't need Jim Nollman to tell them that dolphins were responsive to sounds in the water. With the whales gone and the dolphins starting to burgeon on a fishing ground that had previously been off-limits to them, the men on the boats used their historic weapon against dolphins—the *tsukimbo,* pipes that are hit with a hammer underwater to make a scary sound. And the dolphins *were* scared—at first.

They withdrew thoughtfully, observed the fishermen's habits for a while, and soon detected what the hammering on the pipe meant: *We're sitting on a school of fish—let's keep those dolphins away.* And so, as Jim put it, what began as a scare tactic had become a dolphin dinnerbell. Hammer on the *tsukimbo* and the dolphin came running. The comedy—which soon enough became a comedy of terror—had started. (*The humans signal but the dolphins figure out the trick.*)

The next trick from the acoustical expert's bag—a modern one, often tried—was an underwater recording. Since dolphins were believed to be deathly afraid of killer whales, the sounds of orcas on the hunt were released into the fishing grounds so that the dolphins would flee. Well, dolphins are afraid of orcas, but they aren't afraid of orca recordings. The trick worked for a moment, but then the dolphins determined they were not up against the real thing and they paid no further attention. They continued to steal the squid and hunt the buri with all their slippery aplomb. With a complicated system of nets, the fishermen tried to capture the dolphin as a preliminary to the kill.

The Greenpeace technique is confrontation. Dexter Cate decided he would spoil the fishermen's traps and set the dolphins free.

"I knew what Dexter was planning, and I had mixed feelings about it," Jim told me. "On the one hand, I supported it because they had caught so many dolphins and it was terrible what was happening. But I also believed that freeing a batch of them was not going to solve anything. Dexter would be collared, the freeing would make a lot of news in the United States, and that would be the main effect."

What he couldn't judge was how it might affect his own scheme to intervene between fisherman and porpoise and stall for time while looking for a technique that would actually keep the dolphins at bay as effectively as the killer whales once had.

There was a chance that he was on top of the real solution, and he needed an opportunity to prove it in action. He had determined that a siren sound had an effect on the porpoises, and there was another sound which, if deployed in proximity to the boats, might warn the dolphins to keep clear.

"It was the sound," said Jim, "of dolphins being caught—their own distress signals played back to them."

By constantly talking to the fishermen in an effort to make them amenable to further experiment—although the government's efforts along those lines had failed—the interventionists actually wangled their way onto one of the dolphin roundups they had made it their business to stop. Jim welcomed the chance to show the fishermen that dolphins are more susceptible to sound manipulation than they believed, and he had a chance to show them that the siren sound controlled dolphin behavior more effectively than anything tried earlier. He never had it in mind to trick the fishermen or to participate in destroying the roundup.

Dexter swung into action. He had his wife and son, as well as Jim, with him when they arrived in Katsumoto Village on February 28, 1980. In the account later circulated in a Greenpeace leaflet to raise funds for legal fees, there is this description: "Earlier that day, Dexter and his family watched as 500 dolphins and small whales were hacked to death, drowned, or were left writhing on the beach to be disposed of the next day. . . . Wounded dolphins swam in frenzied circles supported by their companions in a futile effort to keep from drowning in the blood-red waters of the bay. . . . Working alone throughout the night, [Dexter] loosened the ropes holding the nets and struggled in the cold, pulling stranded dolphins back into the water. Those who could swam away to freedom. Too fatigued to return to the main island, Dexter waited on the beach for the fishermen to return."

From the standpoint of the folks on Iki, Dexter was as culpable as somebody who had freed the wolves while cattlemen in Wyoming were up in arms about them. But dolphins, like wolves (the wolves only belatedly), stir men's hearts. They stir even the hearts of some of those who are closest to the killing. So Dexter survived that confrontation with the villagers—it takes a really strong soul to sit and wait for the mob in a situation like that—and he was turned over to the police for prosecution. American courts are slow and Japanese courts are slower, or at least more intermittent. Two days in court per month were Dexter's allotment, so months went by while Greenpeace proclaimed his plight, and the world was as angry as ever with the islanders. For their part, the islanders fumed in continuing indignation—and Jim Nollman found that the time he had spent in wedging an opening for the fishermen to accept his help had been obliterated.

Jim wasn't mad at Dexter Cate. It would be hard not to admire Dexter, for he took his life in his hands. There was no assurance that the islanders, if they became angry enough, wouldn't have done some of their hacking on Dexter. But Jim was sorry to see another chance at using communication techniques with the dolphins go glimmering. In a sense, Dexter had been his mentor on the Iki situation. He had steered Jim toward making acoustical demonstrations in Japan as a way of "keeping the issue hot" so the fishermen wouldn't be able to bring off the massacre without the

Japanese people being aware of what was coming up. Dexter had brought off a useful coup when he found the Department of Fisheries sitting on secret information that mercury samples in dolphins taken for eating were ten times the safety minimum. That was Dexter Cate's best moment, and it stopped dolphin fishing overnight—until the Department of Fisheries decided that dolphin should, as Jim recounts, be caught and frozen while they took a new reading on the mercury levels overall.

This up-and-down battle has continued. After the Japanese had held him a few months to demonstrate that freeing porpoises could not be taken lightly, Dexter wound up back in Hawaii, working for Greenpeace. Jim wondered if the fishermen he had tried to befriend in order to apply new principles of saving dolphins would give up the notion they had conceived about him in the wake of Dexter's night of reprisal on Iki Island.

"I wasn't there to confront the fishermen or even to protest the dolphin massacre," Jim says. "Just to find the sound that would repel the dolphins from the boats so the fishermen wouldn't have the same reason for killing them. I still think that's the solution, and I think we came close to working it out. It wasn't Dexter, it was the $80 a head that blew up what we were working for. But I had worked so closely with the fishermen that when the dolphins were set free, they decided I was Dexter's spy, or maybe the ringleader, and they broke off every contact I had ever had with them."

JIM AND A WOLF NAMED ARCTIC

I don't like to leave Jim there, glooming over the failure to rebalance the situation at Iki. I would rather leave him in the posture that most suits him—crouching and alone, attempting communion with an animal by trying to speak to its very heart.

After the Iki troubles, Jim took a week to howl with a pack of wolves—to become one of the howlers. A white wolf named Arctic, leader of the wolf pack, was the creature he needed to impress in order to succeed in this sing-along. The wolves treated Jim somewhat like the orcas had. They listened to his early efforts and then decided that if he was going to howl, they would teach him to howl better.

"The wolves had different songs for different times of the day," Jim says. "I can't be sure of the content, but as far as the form went, I knew it was the oldest song I'll ever hear. There's a structure. It was very similar to classical Indian music, so there may have been a reachover there of some kind. When the time of day changes, the new song coming up will have a different rhythm and scale. The idea that if you've heard one howl you've heard 'em all is false.

"The wolves were the best musicians I've ever played with, better than the whales. I felt with the whales that there was more of a dialogue, more potential for a collaboration; but the wolves did amazing things. They would resolve chords and do other musical things that were very, very complex. I guess you could say there were two packs, for they were divided, about 400 yards apart, and when the first pack started, the second would follow, and they were communicating. Something in the howls went through.

"Music isn't static; it's a transfer of thought. I've been a musician all my life; it's my way of getting through. With the animals I've met, sometimes I know very clearly that this is their way of getting through. What the wolves have, when they sing, is as deep a conversation as people in ensemble have. You shouldn't underestimate that."

The alpha wolf Arctic, leader of the pack—or maybe the songleader, for leadership in a wolf pack sometimes changes with the job—took a long time to accept Jim, but then he did and they produced harmonies together. Woe unto Jim when he broke the harmony. Arctic would stop. The pack would stop. They would pace. And they'd give him another chance.

After six days of night-and-day howling—which included ecstasy and screaming at dawn when the wolves seem to come to a sobbing climax of some kind—Jim stepped out of the ensemble. He has since claimed that if he howled with them for five years, he would know for sure what they were saying. But Jim was looking to make a date with some howler monkeys in Panama.

Most of this, the successes and the failures, are going down on tape. If it is never translated in its entirety, it will still provide evidence that it is very wrong to think that man, and only man, becomes drunk with his visions. The wolf who cries at dawn and bays his triumphs by day is clearly having visions, and Jim knows, from being taught to find the right note by his shaggy friend, that you must howl with visionary skill if you expect to find room in the pack.

15

Some Disputes About the Right to Play God

WE HAVE FOLLOWED IT ALMOST TO WHERE IT LIES AT the present moment—the quest for interspecies communication. We are almost at the end of the story now, admittedly an incomplete one. One significant part of the tale remains to be told.

An ocean diver named Steve Sipman, whose own exploits are part of this, deplored certain aspects of the search for animal communion, though he became as affected by that search as anyone alive. He came to hate the route by which intercommunication was approached—keeping porpoises imprisoned in small tanks. He swam with the great whales but then he saw that this made too many others want to do the same. Experienced divers feared calamities when hydrofoil tour boats began rushing into the whale breeding grounds. These boats are potentially dangerous because the hydrofoil system seemed to leave the whales unalarmed. Off the coast of South America, a calamity did occur when a hydrofoil smacked into a whale, killing a stewardess on the boat and probably the whale, too. It proved the divers' fears were not ephemeral.

Steve Sipman called the great stampede for animal communion the Whale Rush. He loved whales, hated the Whale Rush, and became the author of an attempt to remind humans searching for interspecies communication that they have no right to grievously injure the animals they wish to communicate with. This chapter is, in part, the story of the "porpoise revolt" managed by Steve and carried out with the help of apprentice scientists who, in swimming with whales, had changed their minds about the nature of the world and the rights of scientists. But the story really begins twenty years earlier, in the great curiosity the human race brought to examining races of the sea once the aqualung made an underwater explorer out of every person who wanted to be one.

From the beginning of the skindiving era, divers have grabbed giant sea turtles and let them swing them away on a wild ride. In Pompano Beach, Florida, where I lived in the late 1950s, skindivers often came out of the ocean holding off a shark they had teased with the rod of their harpoon gun. Under the sea, familiarity led to impudence; impudence to recklessness; recklessness, succeeded with, gave way to the firm idea that humans need have no fear or awe of any creature down there.

What were the explorer-adventurers like who took skindiving toward its most awesome moments? Some were like Jim Nollman, but a few were like that stormy gentleman I met who was so sure he understood the habits of creatures of the ocean that he refused to be alarmed when a great white shark turned up as his swimming companion. Fred Baldasare was an underwater swimming champion and such an incredible physical specimen that he literally glistened with physicality—the strength seemed to be popping from his skin. The ocean divers who run real risks are often like supersurfers. Fred, though older than others and white-haired, was daddy of the breed. In 1962, he became the first channel swimmer ever to make it underwater from France to England (time: eighteen hours, one minute). Looking like a handsomer Hopalong Cassidy and convinced that he could deal with sharks the way Hoppy could deal with a herd of cows, Baldasare claimed a knowledge of ocean communication habits that put all professors on this topic to shame.

He scoffed at theories on how to deflect a shark attack. "If the shark wants you, he gets you," he said fatalistically. His point was that sharks generally don't want you and that, in any case, they take long enough to make up their minds that someone as shrewd as himself has time to withdraw and live for another swim. Baldasare said he could tell by the sharks' body movements when they were getting serious and it was time to leave the area. He did not trouble himself about sharks who merely cruised about or made what he seemed to consider playful runs at him.

"I've had sharks dive-bombing me, but I don't consider that an attack," he said. To him, it was like a dog rushing the postman for the fun of it. If you knew how to keep your cool, the shark was as easy to handle as an energetic dog. Trumpeting his plans for swimming from Cuba to Miami Beach through shark-infested waters (the swim was called off but apparently because of currents, not sharks), Baldasare described a world alien to most of us—but becoming familiar to the skindiver.

No other skindiver, though—and Florida was full of divers who would glide among sharks and barracuda with no sense that they were risking their lives—had anything to compare with Baldasare's tale of his swim with the great white shark. It happened, he said, during an eerie swim under gloomy weather conditions with a Navy escort vessel alongside to pull him out in case he tired before reaching shore. I've forgotten now what he hoped to accomplish with the swim or where it was made. That was the trouble with Americans, Baldasare said—in Europe he was "better known than Mickey Mantle," but Americans couldn't see the excitement of what was going on in the ocean. They didn't know who the real athletes were.

While he made his swim, he claimed, a great white shark had joined him and they traveled along together. He had no fear because he could read the shark's movements and would have known in time to signal his Navy escort if the shark turned bad. His escorts finally saw the shark and insisted on pulling Baldasare out over his protests.

Malarkey? I have a hunch it wasn't. In any case, Baldasare's report about being able to tell a shark's intentions was comparable to what all the more experienced skindivers said—and the reason they were able to explore the Florida reefs with little qualms about sharks though the sharks were there in great numbers. Baldasare, all nerve and gusto and treating the ocean as a kind of Roman arena where the important thing was to be the champion, was like a climax to the first phase of underwater swimming.

Skindiving was the ideal medium for many others besides those with fish to stab and records to set. Cousteau's voyages on the *Calypso* set the stage for a movement quite different from his own lonely, observant journeys. Inspired with the dream that they could actually make communion with creatures from another world, adventurers turned up who wanted to talk with whales, and they wanted to do it now—no waiting around. Many did not even take the trouble to learn skindiving. The mere appearance of a whale near a small boat was enough to propel them toward ecstasy, and the would-be communers with whales appeared in great numbers. The following is reasonably typical of their journeys.

THE SEARCH FOR WHALESPEAK NOW

Off the coast of Baja California in the mid-1970s, a self-described group of ocean nomads sailed off in search of the great whales. Joel Andrews took his harp. Paul Winter, who, like Jim Nollman, became one of the best-known musical communicators to come out of the Whalespeak Now! searches, had a saxophone to help him mirror the sounds he heard. David Dowling took a cello.

A filmmaker named Will Janis believed the musicmakers could rouse the consciousness of the whale, and he had chosen a truly symbolic occasion for making the try: Valentine's Day. Altogether there were seventy in the party, which made a 1,300-mile trip out of San Francisco in the hope of having an ocean rendezvous with whales. They didn't expect conversation, but they did expect communion—a sense that, through music, the whales would see that they were trying to convey that they respected this other civilization in the sea, wanted to be in touch with it, hoped the whales would give a sign that they understood.

"We set up our tents in the sand dunes overlooking the channel, shed our clothes, and began to live out our fantasies of being on a desert island," wrote James Ricketson, one of the participants, in a newspaper account. "We were probably as far away from civilization and its attendant evils as it is possible to be in the latter half of the twentieth century."

The floating concert started early on Valentine's Day. From a raft in the middle of the channel, where it was believed that whales migrating to Baja could be intercepted, a multicolored balloon danced overhead, catching shots of whales and whales communers from on high. Ricketson was disappointed at the lack of interest on the part of the whales who passed near them. Swimming in pairs, the whales took no notice of the tootling—no matter how strenuously it was performed—and passed down the channel.

Then, in an experience very similar to Nollman's in causing far-off porpoises to perk up when he hit the right note, the whale communers switched to "Amazing Grace," and seemed to have found the combination. From the several mothers and calves who had been swimming near them, a mother and calf ducked out and angled themselves in perfect unison toward the source of the music. As mother and baby came closer, making bubbly music of their own by expelling through their blowholes, the musicians on the raft went from "Amazing Grace" to "Ocean Dream," the song that expressed their full tribal yearnings:

Ocean child
Come now home,
Holy wonder,
Holy one . . .

"And then," wrote Ricketson, "as if in appreciation of the music or at least to acknowledge that she recognized our presence, the mother surfaced in front of the musicians, her enormous shiny back rolling out of the water in a seemingly never-ending arc, cleared her blowhole, and sent a fine-spray fountain of mist into the air that caught the sun's rays and produced a rainbow. Everybody on the raft let out whoops and shouts of joy."

Ricketson found that "some people on the island felt that interspecies communication did occur. Others felt that it only occurred in our imaginations. Perhaps it doesn't matter. If the communication that was felt to be taking place is an illusion, it is a much healthier one than the collective Western illusion that separates man from other life forms and views animals merely as things to be eaten, as a source of leather and fur, or as objects to be locked in cages and admired. . . . It is not enough merely to be intellectually aware of the interconnections of all life but to know it by experiencing it."

THE WHALE SPEAKS—AND SAYS, "GO AWAY"

One of the most ambitious of many journeys into whale country was led by Ron Storro-Patterson of El Cerrito, California, who seemed to me the best-versed and the most meticulously observant of a huge group of biologists, naturalists, and researchers who, in taking a new look at whales, were discovering odd traits and capabilities that the whalers—always in a rush to kill—had managed to miss. Ron's most particular quarry was the great blue whale, twice the size of the largest dinosaur that ever lived, known in sizes up to 120 feet. Swimming, sailing, and flying, Ron zoomed in on the blue whales wherever he could find them.

One day he saw a sight that no one in the history of whaling had ever described—a creature that appeared to be twice the size of the great blue whale, which is said to be the largest living thing earth has known. What he saw was a great blue whale with its stomach extended from its throat for eating purposes, and it could gulp this back inside again. It was an unbelievable feeding pattern, but Ron photographed it so that the world would know what a great blue whale is like at lunchtime.

Out of his several forays among the blue whales came what he calls "Blue Whale Conversations," tapes that are far stranger, if not more esthetically exciting, than the humpback tapes collected by Roger Payne. Ron advised those seeking the recording—either out of curiosity or in the belief they might be able to run it through a computer and make some headway in translating it—that they would have to listen through headphones, turn the treble down or off, and try to *feel* the whale as much as hear it. "You have probably never heard anything like this," he said, "because it is so low, approximately 20 hertz, or just at the threshold of our hearing. Many people compare listening to the blue whales to the experience of sensing an earthquake! The vibrations count as much as the sound."

Inability to penetrate the whales' verbal patterns hasn't stopped whale explorers like Ron from gaining an appreciation of the whale's gesture system. Although he organizes the journeys that take romantic Whalespeak Now! supporters directly to the ocean playgrounds of the great blues, Ron himself is as rigorously scientific as anyone in the field. He has also had some of the most extraordinary experiences of any of the whale divers.

In the Sea of Cortez, on the lower Pacific Coast, in that great whale gathering ground known to whalers, Ron and another diver were in a small boat when they saw "a wall of water" coming toward them. They just had time to glimpse that behind this wall of water, pushing it—as an arm might make a wave, swooshing it across a bathtub—was a great gray whale.

"Chin above the water, coming right at us," said Ron. "The gray whale can swim fast. He was up to about 10 knots, building a bow-wave like a ship. I guess it was frightening if we'd had time to be frightened. It was fast. Almost as soon as we can see what's happening, the boat is riding to the top of the wave. The whale dives and we're perfectly okay, we're all right. But the whale is telling us, clear as can be, 'I'm letting you off this time; I don't want you here.' And we got out."

All the divers who worked closely with whales continually questioned what Dr. Lilly had called the "ethic" of the whales. When humans molested and massacred them, why didn't whales use their great power for retaliatory action? Were they too dumb to think of it? Or, at bottom, were they too basically kind?

A reasonably definitive answer to that question has long existed. It could be deduced from a strange encounter of the steamer *Seminole*, sailing from New York to Jacksonville, Florida, in 1896. For *Sea Front* magazine, writer Frederick P. Schmitt reconstructed those few occasions when whales have chosen to pit themselves against human intruders. The *Seminole* incident is perhaps the most striking.

As Schmitt recounted the episode, a band of what are now believed to have been sperm whales (at the time, no one was sure of their identity) had a run-in with the steamer as it came out of Sandy Hook. Half a dozen of the whales surfaced suddenly at the *Seminole*'s bow and then the ship plowed into one of them, "causing billows of bloody spume to explode from the mortally wounded whale's spout." Schmitt's report continued:

> The shock reverberated throughout the ship. Crewmen and passengers dashed to the rails for a glimpse of what was happening. The quiet summer air was electric. Meanwhile, the remaining whales appeared to retreat a short distance, when unexpectedly they turned and, with great, broad flukes thrashing madly, lunged directly at the ship. Again the *Seminole* trembled from stem to stern. The first pair of encounters knocked her sideways off course, spilling the curious onlookers to the decks. Suddenly, the whales sounded into the depths. For an instant it appeared they were gone. But again they attacked, with renewed vigor, loosening steel

hull plates and disarranging delicate machinery in the engine room. By now seawater was seeping through the ruptures.

Four times the whales charged. Saloon furniture was broken loose from its fastenings. People were bruised and injured. Women fainted. Finally, their fury spent, the sixty-foot leviathans were wounded and exhausted, and the captain was able to pull ahead of them by laying on steam. It was a close call to be sure.

The *Seminole* episode, plus a handful of comparable incidents, prove that aroused whales can stage a tantrum of smoking proportions. Whales are virtually never aroused, however, and remain benevolent for decades on end, overlooking the risky and often presumptuous undertakings of the fly-creatures who attempt to penetrate the whales' private world.

The divers who tried for communion in the whales' underwater habitat were well aware that they survived the whale swims only because the whales took special care that the divers weren't crushed in the process. Ron Storro-Patterson described to me the whales' habit of scratching their barnacles on each other. If they had done this to a diver, it would have left him like a small white radish in the hands of a vegetable grater. What would happen if a whale really chose to hit a diver, or a boat, with force?

"If a whale hit you with force," Ron said, "it would be a little fatal. You'd get a hole knocked in you. The flukes are where the power is. They're a rather incredible instrument. In the phenomenon of breaching, when the whale is taking itself almost clear of the water, the flukes will be whistling forty to fifty tons of whale right into the air."

The secret of the whale swims, Ron said, was the whales' willingness to let them happen. "We always talk about a diver who has caught up with some whales. Well, you're in the whale's element. The human diver is totally awkward compared to the whale. He's at the whale's mercy. We say a diver 'approaches the whale,' but even if the diver is swimming toward the whale, the whale is really approaching the diver. He's allowing it. He's setting up a happening by not swimming away."

THE "LUMINOUS COMMUNION" SPARKS A REBELLION

A special benefit of working with Dr. Lou Herman of the University of Hawaii on his language experiments with captive porpoises and in ocean observation teams was the chance to experience what only a tiny handful of adventurers had ever experienced up close—a whale in its habitat. They had boats, airplanes, and skin-diving gear for close-up approaches to the humpbacks in their nursery waters. Before whale diving became a destructive fad, they collected bushels of notes for Herman on the whales' habits. Herman was intense, somewhat methodical, shrewd at attracting some of the cleverest and most indefatigable of the advanced science students to work with him. They could be as intense as he was—more so.

Two of the students who worked with Herman were Steve Sipman and Kenny LeVasseur, and later I met them both, obtaining their versions of what has been called the "Hawaii Liberation Case." Sipman was a brooding Viking, who could have led dragon ships in the Middle Ages. Totally articulate when he chose to be, he was elusive and hard to know—impatient with newsmen and interviewers who look for "an angle" instead of wanting to know the complexities of a situation. Kenny LeVasseur was a college-boy idea man, popping up with notions for an underwater wet suit that he thought would make the ordinary swimmer faster than the fastest swimmer alive. (He came from a family of expert swimmers.) Sipman and LeVasseur were both extraordinary, as was Herman, and it set the stage for what would happen.

A few years before whale diving became a familiar Hawaiian adventure, Ken Norris had prodded a Honolulu machine-shop operator, Jimmie Okudara, into creating a craft which was dubbed "Norris's Nausea Machine" or the "Semisubmersible Seasick Machine." A mobile observation chamber with a plexiglass bubble, it could zip through the ocean while providing a 360-degree view of the whole great panoply of the reef.

"The hydrobatics of porpoises are nothing short of incredible," Norris had written in *The Porpoise Watcher*. "We once watched an underwater arabesque including six or eight animals that twisted in a complicated spiral as the animals rose through the water. Another time, while observing rough-tooth porpoises far out in the bucking sea twenty miles from Oahu, I watched in amazement as a pair of these awkward-looking animals swam repeatedly by me. Every visible nuance of motion of one was reflected instantly in the other. They swam as a team, not outdone by the Blue Angels."

What Steve Sipman saw, from a plexiglass bubble zipping through the sea, was stranger still. "Wild dolphins ride the bow," he told me later. "Then you see them making trails in the water like a skywriter—only these are feces trails, and it is carried out with such spectacle that you begin to suspect that it's a design they're making and they're very conscious of the design and of seeing how far they can go. For all I know, it's calligraphy done with feces. Because the mind of the dolphin is strange and it's busy. It's always busy. You watch dolphins in the sandy bottom and they're drawing in the sand."

The view from the bubble was astonishing, but it kept the humans discreetly separate from their subjects so that they were watching a performance instead of being absorbed into the drama. With the whale swim came absorption and a radical change in how Steve and Kenny felt about toying with the mentality of captive porpoises.

Tumbling from a kayak, plunging from a boat, Steve and Kenny found the whales were easier to approach in groups than when alone even though there might seem a chance for a giant squeeze-play—with the human locked between the whales. Within seconds of meeting them, fear was the last thing on Steve and Kenny's minds. The size is awesome, but it can be sensed immediately that the whale has a total sensitivity all up and down that fifty-foot length—and no plans for turning its power on the diver.

"You are the small bubble, they are the large bubbles," Steve said. "In the water, you all weigh the same. You can get freaky and decide that you *are* the same. You're conscious of the whale, not as pictures you have seen, but as a creature new to you that no picture has ever shown. The humpbacks flow, they *bend*—these great bodies, flowing, bending, it's beyond describing, incredible. You see them hovering, or they rise to greet you, and it is like engines coming up in the train station—*Chooo!*—and now they are there, four at once.

"Sometimes you can see way below you and the whale is fading into the blue, this beautiful blue. Or you'll see the tops of the pectoral fins, which are white. I'd come up to breathe at times, sitting on a surfboard for a while, and there might be a pod of six or seven of them, and they would be here and here and here, all around. The porpoises who stay near the whales might start to play with the calves. And the calf is like a big puppy to them.

"When I was trying to get a good photo of a calf, a male came toward me and shook his head back and forth. That could be just mammalian behavior, something a bull would do. You see the same thing in the dolphins. But I think it was more likely body language. He was saying, 'Get away from the calf.' The mother puts herself between you and the calf. They are protective, but you do not feel threatened."

The humpbacks did not seem surprised to find humans coming for a whale swim, but the humans were astounded. They were the first to try it, Herman and crew, or at least they ran into no one else who had. What they knew about it, they found out by experience—there were no precedents to go by. At the beginning, as they piloted this "luminous communion" with creatures who were sometimes willing to commune, Lou Herman seemed as entranced, Steve claims, as the others. Herman once forgot himself entirely, Steve said, and jumped from the bow of a boat that was going forward.

"The skipper freaked when Lou made his crazy dive, but it just proved how wild we all were to get in there with the whales and find out what we could. I know that Lou was excited, first time I ever saw him truly excited. His eyes were big. It's hard to talk about mystical experiences, and not everybody welcomes them. But I know he had the feeling, because it's very humbling to be next to something as big as a humpback. Stand next to a giant redwood, 200 years old, and it humbles you. You're a big-time researcher and you go out in the ocean and now you're alone except for these forty-foot whales—that's very humbling. Chooo! Chooo! Chooo! Lou felt that as much as we did."

It is not financially safe for a scientist-teacher to be rapturous about an experiment—or to let the experiment grow into something larger than mere experimenting. He has to put down rapture in favor of something more to the point: the soundness, the objectivity, of data.

"Data," Lou Herman had told the divers, "is sacred."

Herman, using student helpers, was eight years deep in the experiments with Puka and Kea. He felt right on the verge of a breakthrough when the dolphins would capitalize on their long training and show high reponsiveness to complicated signal systems he was working out. Steve and Kenny, at least, began to lose interest in Herman's language goals and to view what they considered the dolphins' deteriorating condition with alarm. In Steve's view, Herman was in danger of losing grant money, was having to be overeconomical in the management of the experiment—the experiment was running out of steam, and the dolphins, usually accepting creatures, were running out of patience with their human captors.

Steve thought—Herman does not agree—that the dolphins, shortly before Hermans' students conceived their revolt, were out of hand and already in rebellion themselves.

The dolphins, claimed Steve, were "hurting themselves over and over. They would smash the speakers during an experiment. To see that day in and out wears on you. If they were not self-destructing, they were at least self-abusing. If the computer made a mistake or if the experimenter missed a reward—the porpoise is older now, and it's pissed off.

"Sometimes they'd smash the wall. Sometimes they'd snap at the experimenter. Sometimes, even without an experiment, the dolphin would go into some sort of a rage."

Michael Fox, the animal-rights leader who was trying to reform everyone's outlook on the responsibilities of humans toward the animals they have in thrall, had said something that sounded spooky and entangled but was also unforgettable. "I'm attempting to show," he told Emily Hahn (as reported in her book about the chimpanzees, *Look Who's Talking*), "that man is both animal and God, that there's a biological, evolutionary, ecological kinship with all animals; an inseparable, interpenetrated interdependence, that man's now in the position of God, in terms of responsibility and stewardship."

Once they'd been on the whale swims and felt that they were actually in contact with the whales, Steve and Kenny were all for interpenetration and interdependence. As for man being in the position of God, this was where they decided to draw the line. The porpoises seemed to be approaching a nervous breakdown as a result of their long-term imprisonment. Steve felt that something must be done before the situation grew worse.

Herman's own reading of the situation was entirely different. He told me, in the aftermath of the revolt, that Steve and Kenny were peripheral to his experiment, had little sense of it or where it was going. Steve and Kenny, he suggested to me, were "tank cleaners."

"I wasn't a tank-cleaner," said Steve, "I had been at the laboratory in Hawaii since 1972. I was a student at the University of Hawaii and had taken about four courses from Lou Herman, getting A's in all the courses. I was studying the behavior of the marine mammals. I've spent now I don't know how many years"—he was twenty-nine years old as he spoke and had just gotten married—"studying the scientific information on porpoises. I hated the conditions, the way the tank was, and I was worried about the tension hanging over Lou Herman because he thought the grant money might stop. It's an old sore now, but he put signs all over the laboratory—'No toys for Puka.' That was to get her motivated. He thought if she had no stimulation, then when the experiment came along, at least she'd be glad to take part. 'No swimming with Puka.' It was sort of what you might do with a child." A year earlier, Steve claimed, Herman had told them "never put a dolphin on deprivation," but now the pressure to obtain results was mounting.

Everyone in Hawaii had heard tales of so-called kamikaze dolphins being trained to fight wars of the future. The Pentagon always said this was balderdash, but everyone from Linehan in the *Geographic* to the *Christian Science Monitor* and *Penthouse* had traced bizarre twists on the idea that dolphins were being trained to blow up enemy ships as living torpedos (naturally the porpoise wouldn't know, until blown up along with the ship, the real point of the training).

Steve claims that he and others feared the experiment with Puka and Kea would break up and the porpoises would be handed to the Navy for just such deadly training. Some of Herman's work, he argues, was classified. They were suspicious all around.

Steve brought himself to so big a burn he began to organize a secret campus force to spirit Puka and Kea out of the tanks and into the ocean. By Steve's account, Kenny had a wonderful conscience about dolphins, but he talked too much to everybody; so he wasn't told about the plan until the last minute to insure no hints would reach Herman that something was up. In a meeting of the conspirators, it was decided to call the movement to free the dolphins by the same name as the freedom route for slaves before the Civil War: the Underground Railroad. "That clicked right away," said Steve. "Sort of gave us the old Bicentennial boost." He had eight conspirators at the time, recruited more. Why was he so angry? Didn't they fear prison?

"A lot of people," Steve told me in his dark and ruminative way, "are still in pre-Copernican ideology. In the first chapter of the Bible, God says to Adam and Eve, 'Go for it.' God says, 'I made this all for you'—and people believe that. They believe they have manifest destiny or divine right over all the creatures of the earth."

The plan was to drain the tanks, slide stretchers underneath the porpoises, and take them out in the middle of the night. Steve calculated he would need four bearers per stretcher. They also wanted to announce the liberation with huzzahs—before the theft of the porpoises, they made plans to issue press releases the following morning. In some ways, the great porpoise snatch smacked of the jewel-robbery films all had seen. They had lookouts to keep an eye on the harbor police. They wore white lab coats and face masks so, if police stopped them, they could answer, "Please stand back. Dolphins are very sensitive to human germs. We have to move them at night to keep them from being infected."

"When we loaded the dolphins into the vans, some people saw us," Steve said. "It was about three in the morning, and you'd think they'd wonder why we were putting dolphins in vans at that hour, but I guess the abnormality of it made it seem normal."

Four or five police cars came toward them, didn't stop them; they never had to give their fancy alibi a try. At the usually deserted Kaena Point on Uokohama Bay, where the release was made, the beach was dotted with overnight campers. The kidnapers had forgotten it was Memorial Day weekend. But the campers paid no attention as the vans were unloaded and the dolphins were put into the ocean after eight years in the tank.

"Dark, no lights, beautiful moon, clear sky—the water was kind of inky black," said Steve. "Slosh slosh in the water and they were gone. We sat around, sort of relieved. Some of the people went home and worried. Some went home and applied their typewriters to writing press releases."

On a blackboard back at the tanks was scrawled the porpoise burglars' parting message for Lou Herman: "Went surfin'—Kenny, Puka, Steve, and Kea. Aloha."

Kenny was a necessary part of the porpoise pullout because, like Steve, he slept in a room near the porpoise tank and there could have been no theft without his cooperation. But no one had anticipated that Kenny would dominate the news about the theft the next morning, discussing the case with reporters and giving quotes that sent many an intended interspecies communicator reeling.

How had the plot developed? Kenny said that he'd had a telepathic conversation with the porpoises, who told him to free them. In other words, the *porpoises* had instigated the kidnaping. When I interviewed Kenny while he was awaiting appeal of the six-month jail sentence that had been imposed on him, I found him to be the voluble, excitable, wild-dreamer type that Steve Sipman would later describe to me.

Kenny was convicted of conspiracy but Steve, as I write this, is still awaiting trial. Nobody seems intent on clapping them in jail while the legal struggle drags on. They had expected a testimonial on the need for the kind of act they had committed from Dr. Lilly, but he declined to assert that it was fair play to free Lou Herman's porpoises. Ken Norris was asked if he would testify for them in court, but, according to Steve, Norris said, "I'll be a witness—for the prosecution." Steve and Kenny had succeeded in doing what no one else had ever done: they put Dr. Lilly and Dr. Norris, for once, on the same side.

Steve told me he had no plans for going to jail. A cage or a tank, he said, is not the place for an intelligent creature, "and I'm an intelligent creature."

"LET'S NOT HAVE TALKING AQUATIC SERVANTS"

In the porpoise labs and in the press, as well as in court, the conspiracy failed. Except for the rainbow warriors of Greenpeace and Project Jonah's Joan McIntyre, few who were involved in language experimentation or in the animal-rights cause believed that extralegal liberation was the way to relieve the plight of the porpoises. Many continued to believe that "proving the creature has language" would very likely be the only way to afford porpoises the freedom that no human institutions will grant them so long as they are considered animals like other animals.

Steve Sipman was one of my own greatest surprises in trying to reach the various persons who have been affected in one way or another by the many-fronted struggle to produce evidence of interspecies communication. Watchful behind a reddish brown beard and thick mustache, he has an air of intensity and shadowy drive that make him seem to me the prototype of all those divers who already felt metamorphosed by their experience into another kind of human being—those who were Born Again by finding how different it truly is down there in the sea.

From his own standpoint, he wasn't a wild liberator, he was the cautious voice of returning sanity. Nobody, in the Hawaii Liberation Case, has ever recanted his position. Herman is still mad at the divers; Steve Sipman is still mad at humans he considers to be exploitive and ruinous toward creatures they don't understand.

Steve told me, darkly, "One of the big causes of death of dolphins in captivity is pulmonary disease. Our effect on the dolphins is something like our effect on the Hawaiians. The dolphins know body language, and they're smart enough to abstract. There's an excellent chance we *will* teach them an artificial language, some language we can both feel at home in. And then it's a question of what happens from that. It could be like the cotton gin and the blacks, couldn't it? *Look, the fellow's clever, he understands us; get him on the old cotton gin.*"

In their bedrooms near the tanks, Steve said, he and Kenny "could lay in bed at night and hear the dolphins making sounds. If they were distressed, you'd hear them making sounds. You get familiar with the dolphins' moods. Sometimes loneliness. Sometimes—restless. You know what their environment is. You know what they're missing. Sexual frustration. You understand this. The dolphins don't consider themselves anybody's pet.

"Nobody at the facility could avoid noticing the dolphins' ability to deal with abstract terms. They can *deal.* They'll deal ten times better than we ever thought they could. And this is dangerous for them because, given the mood of man, they'll be made into aquatic servants, most likely. You know how the Navy worked to turn them into self-guided torpedoes. You say you'll train them, but training them, and even trying to talk with them, becomes a prelude to their destruction. And we say this is what makes the experiments worthwhile; it's what man has to accomplish.

"And, as it happens, I differ. I differ."

The darkness of his mood subsides again, and he begins to talk about the whales he has known personally.

Somehow he wanted to work his way back, he said, to being an underwater investigator and naturalist so that he would be able to explain it all a little better than anyone has so far been able to. He talked about the massive note taking that had been done for Lou Herman and said, "I wasn't always sure if I was the observer out there or the observed"—when he was in the kayak, that is, and the whales would pop up for a look. "Maybe the whales took more notes on me than I did on them. You'd see the big eyeball looking at you, and there's no question that in some ways the whale is more aware than you and I are. He's not a goof.

"The whales seem unpredictable until you've watched them a long, long time. And then they seem predictable—until you watch them even longer."

LOU HERMAN: A NEW LEASE ON LANGUAGE

While Kenny LeVasseur was founding a new group called FREED (Foundation for the Release of Every Enslaved Dolphin), Lou Herman began his new dolphin language experiment with two animals, Phoenix and Akeakamai. In Hawaiian *akeakamai* means "lover of wisdom."

The work has proceeded speedily. Phoneix and Akeakamai have taken language lessons to a point far beyond the twelve sonic tones learned by their kidnaped predecessors. Nouns, verbs, and adjectives are part of their new vocabulary, which is headed toward the fifty words once predicted for Puka and Kea but could go much higher.

All sorts of objects may be in the tank with the dolphins at any given moment, but they know the fountain from the Frisbee, the ball from the hoop, and they know when they are being asked to fetch something, leap over something, or perform some other gyration. If the dolphins show the same aptitude in the language lessons that they have in stunt training, this new Herman experiment could open the way toward porpoisespeak on a developed scale.

In a 1979 issue of *People* magazine, Herman was quoted as saying that the progress in the new run of experiments has exposed "a lot of myths about the uniqueness of man." He was sounding very much like the interspeciesists. When I had spoken with him only a short time before, he had still been mulling over the question of a dolphin's ability to master abstract concepts, and had doubted that the dolphin could go very far.

It would be possible to argue that the conspirators got a part of what they wanted from the porpoise kidnaping—large-scale public soul-searching on the moral problem of keeping sensitive, intelligent, language-using animals in prison—and that Lou Herman got something from it, too. His newest experiment is better funded and is receiving better notices than the original. It seems more ambitious. Earl Murchison, Jayne Rodriguez at Flipper's Sea School, and many others cite Lou Herman as the researcher who is providing growing documentation that dolphins are beyond the trick stage and that they are able to have verbal exchanges with humans that come closer each day to expressing something more complex than *fetch* and *jump*.

Herman was bitter when I talked to him soon after the porpoise theft. Since then, with the new experiment going well, his ambitions have soared. Now he has a scheme to try language exchanges with free-swimming whales. "Our goal," he told an interviewer from *People* magazine, "is to establish two-way communication, admittedly on a very elementary level. We're not going to talk philosophy [with the whales]. They'll request and we'll provide their requests. They'll be in utopia."

Earl Murchison, who has his own operations nearby and, as head of the organization of dolphin trainers, should be in a position to judge said what many others did: "Lou Herman is doing the best work that's going on. It's quite impressive, and we'll have to be around for a while to see how impressive it is."

But there's a point that Murchison won't yield on, even so. He doesn't believe that porpoises will give up their secrets in the wholesale way the interspeciesists hope they will, or that they'll tell their life stories to humans. "That's naive," he said. "You're talking to a creature from an entirely different universe. And this creature has an entirely different perspective on the universe—*entirely* different."

Why should we expect a full interchange? he asks. "How presumptuous! How anthropocentric! What in the heck do they have to say to us, anyway? And if they did say it, how would we understand it?"

Murchison is not an interspeciesist. He doesn't believe an interspecies language is on its way—the kind of language that jumps from "Fetch the hoop" to asking a dolphin "How do you feel about the International Whaling Commission?" Paradoxically enough, it's the unbeliever Earl Murchison whose most advanced work suggests, with precision, why the imaginative theorist in interspecies communication believes the porpoise will soon be outshining all others in the race to find an animal who will one day be on a conversational plane with ourselves.

THE DOLPHINS AND THE SHELL GAME

If we follow Murchison stage by stage in his experiment with Kae the yes-and-no-dolphin, we can understand why he argues there is a "magicalness" in dolphins that is more mysterious, yet more provable, than a facility with human language.

On a typical day at Kaneohe near Oahu, the testing ground for porpoises operated by the Naval Ocean Systems Center, Kae would choose between red and blue "paddles" in response to a series of questions put to her by Murchison. Red was for yes, blue was for no. The questions were always about objects dropped in the ocean.

"Is there anything out there?"

The trainer speaks his question in English so that humans will understand it, but it is represented to the dolphin as an electronic tone. If the drop has been faked and nothing has been put in the ocean, Kae reports this faithfully by choosing the blue (*no*) paddle. When an object is actually dropped, the question "Is anything out there?" is followed by another: "Is it a cylinder?"

"Yes."

"Is it stationary?"

"Yes." (They have it suspended.)

"Is the next one stationary?"

"No." (They have let it go.)

Whether the object is aluminum or wood, brass or steel, long or short, big or small, she touches the correct color to answer the questions. When she says no, she sometimes "snorts a bubble from her blowhole" as though to give the rejection emphasis.

All this seems amazing enough. Who would believe that a dolphin can be readily taught to distinguish any and all varieties of what humans call cylinders? But the part of the experiment that tends to boggle the mind of the spectator—or the trainer—comes after this. Kae can perform the cylinder-identification test just as reliably when the cylinders are dropped, in murky water, well beyond her visual range. She is now identifying the objects through her acoustical system alone. Sight does not play a part.

This is a follow-up to experiments by Ken Norris, who had shown that a blindfolded porpoise can navigate, catch fish, go after objects with just as much skill as if it could see. The ability of the dolphin to perform while blindfolded (comparable to a power that had been already discovered in bats) baffles humans, and yet we have a touch of the power ourselves. As an adjunct to his studies on the marine mammals, Thomas Poulter noted that human blind subjects "detect targets and make size (and, to a remarkable degree, shape and texture) discriminations, by using a hiss or whistle, either continuous or in short bursts, or by pronouncing words with *F* and *S* sounds in them, and some use a clicking noise with their tongues." Once it was understood that this secret sense of the bat and the dolphin can have utility for the human blind, a New Zealand inventor, Professor Leslie Kay of the University of Canterbury, developed ultrasonic eyeglasses for the blind that create the same effect artificially, enabling someone wearing the glasses to detect objects from as much as twenty feet away.

Although we readily accept that humans could learn about, and then adapt, this very specialized sense of bat and dolphin, there is tremendous emotional resistance to the idea that a dolphin could learn the syntactical devices of which humans are so proud.

Any dolphin, though, can easily outclass the most sensitive human in making long-distance perceptions that do not depend on sight. That's what Murchison means by a "magicalness" that outruns what the language theorists are trying to prove. He accepts the view that animals have all sorts of wondrous capacities except the verbal—and fights the Lillyites who see a lingual slant to his own experiments with Kae. If we look at this experiment, though, it is possible to speculate that Earl Murchison is actually laying the groundwork for "teaching the dolphin language" on a very large scale.

Murchison told me he had trained Kae to identify cylinders by using the standard reward system. Fine Columbia smelt were tossed to her if she pushed the right paddle. The same system would work with a human, he said, suggesting that he could use a package of M&M's to get me to make the "right choice" if he wanted me to nudge a paddle indicating the meaning of some word in, let's say, Hottentot.

"We use total and complete positive reinforcement," Murchison stated "When it's established with one cylinder that there's a reward for making the right responses, then I put in a totally different cylinder. At first there's confusion but I lead the dolphin through it. Finally, I put a noncylinder in and very quickly I reward her for *not* touching anything. This is very fast. I don't even give her a chance to touch it. It takes hundreds and hundreds and hundreds of trials, but it gets to the point where you can put in a cylinder never seen before and the dolphin makes the connection.

"At first, the dolphin is just memorizing target objects it has seen before. But you get beyond that. And there's a point where the porpoise says, 'Aha!' "

It is within that ability of the porpoise to say "Aha!"—to

grasp the principle that relates one cylinder, any cylinder, to all cylinders—which prompts some experimentors to want to take a direction Murchison himself does not believe in. Why couldn't the dolphin go any distance? Why couldn't there be an *unlimited* number of "Aha's"?

This is where Murchison draws away. He believes there are limits. He has worked with dolphins for years. He doesn't see them as making an unlimited number of "Ahas!"

Theoretically, though, if a concept as difficult as "cylinder" can be thoroughly taught, then whole classes of similar "absolutes" could also be taught, and some students of dolphin lore would like to see these experiments continued with many dolphins over many generations, eventually taking in hundreds of human words and human ideas. Suppose, however, that an attempt were made to teach the dolphin what lies at the heart of human language—not just absolutes like "cylinders" but nonabsolutes like a *dented* cylinder, a *crumpled* cylinder, a red diamond, a queen of diamonds, a baseball diamond, a diamond ring. Wouldn't the dolphin get lost in our language forever? How do we ourselves figure out such a complicated system? The mass of human language doesn't deal with geometric-mathematic absolutes but with subjective impressions. Could Kae learn the word "boat" as readily as she learned cylinder if we showed her that "boat" includes skiffs, sailboats, tugs, liners, canoes, catamarans, and on and on through several dozen varieties? Could the dolphin go on (and if so, how?) to master the additional, interchangeable ideas: *pretty* boats, *big* boats, *sinking* boats, *blue* boats, *junk* boats, *repaired* boats—and on through hundreds of special conditions?

Isn't it all too much of a maze?

Many scientists believe it is and quake at the thought of conveying language on so vast a scale. But there are interspeciesists who believe that if the right format can be found, dolphins will show ability in grasping such multitudinous concepts just as I—though a Hottentot might think me unable to learn his language—will do very well when guided along by a reward of M&M's for my right answers.

An interspeciesist could argue that the Navy and Murchison are failing to recognize just how illuminating the experiments with Kae, and various comparable tests, really are. Lillyites know how readily dolphins and whales adapt to complicated situations and proceed from this to an assumption that these animals *do* understand many propositions that have to do with judgment (a *good* boat, a *bad* boat). One of the chief reasons the interspeciesists feel they can make this assumption with assurance is because dolphins will snack on certain mammals (*a good*) but won't snack on humans (*a bad*). Since the dolphins are making judgments not as a matter of teaching—we *show* our dogs that biting is "bad"—but as a matter of personal reasoning, or instruction from dolphin parents, we can deduce *that they intellectualize abstract questions.*

Now they may intellectualize differently than humans do, but the difference is more in the conclusions than the methods. The humans in the tuna fleet reasoned: "It is all right to kill dolphins when necessary because we need a rich load of tuna and nothing should stand in the way." The dolphin, although it can and sometimes will take on a shark, seems to think: "The *shark* can be injured or killed but the *two-legs* are different, and I can't use my power on them." This is not a fanciful projection of the dolphin's mind-set but is supported by thousands of instances reported by people all over the globe in all sorts of different circumstances showing that dolphins virtually never attack we two-leg humans.

It indicates, just as Dr. Lilly kept suggesting, that the dolphin reasons and, like humans, has a moral code.

To Murchison's question—"What the heck would dolphins have to say to us if they *could* talk?"—the interspeciesist could answer that the dolphin would continually surprise us with its observations. Murchison was finding, from the Kae experiments, that a dolphin could, to some unknown extent, read not only the exterior but even the interior of an object. Ingrid Kang was sure that this included an ability to read whether another dolphin was pregnant or not. A dolphin who could observe to a human, "I don't want to shock you, Bethie, but you have conceived a baby, dear," would not find it difficult to have something to say that would interest humans. As science-fictionish as this may sound, there is every reason to believe that dolphins can in fact read the inside of another creature's body. The Swedish writers Karl-Erik Fichtelius and Sverre Sjölander put it like this: they said it appears that dolphins can "hear" each other's intestines. Now I do not wish to hear anyone else's intestines, and do not envy the dolphins this particular power, but I know, if I were a Navy officer charged with protecting some great port like New York or Singapore, I might take an interest, even an exploitive interest, in discovering if dolphins can be used to scan the insides of peaceful-looking boats to see if they have atomic weapons or caches of arms in the hold.

The proof that all this is not mere speculation came in an odd way—when Earl Murchison asked Ingrid Kang to use her Sea Life Park dolphins to follow up on a particular idea he had. He had a hunch that, while humans continue to be bamboozled by that old carnival stunt, the shell game, it should actually pose no problem to dolphins.

"The purpose of that old game," Murchison wrote, "is to guess which of three walnut shells a small pea has been placed under." Ingrid Kang set up the shell game with large shells and large black peas—the peas invisible to humans when under the shells. After experimenting with different materials to see what the dolphin could read best, Ingrid soon had a dolphin picking the pea from the shell that concealed it. But the dolphin didn't signal in a rush and dash off, yipping for a fish. In Ingrid Kang's description of the way the dolphin solves the shell game, the most impressive factor was the way the dolphin nosed from shell to shell, carefully considering, nosing again, going back, and at last deciding.

That careful inspection indicates that the dolphin understands the game for what it is: a challenge. It doesn't make idle guesses; it settles down to a very determined effort to *make the right answer every time.* My own guess is that the dolphin understands the game very well. It sees that the fun of it is guessing better than anybody else. And because the dolphin has those seemingly magical qualities that we humans have not developed to the same extent, it really doesn't have to guess at all—it just beams in and gives us the answer in a gesture that humans can understand.

IN FEAR OF PIDGIN WHALE

I reckon I got to light out for the Territory, because Aunt Sally she's going to adopt me and civilize me and I can't stand it.
—Mark Twain, *Huckleberry Finn*, 1885

The scientist does not study nature because it is useful; he studies it because he delights in it, and he delights in it because it is beautiful.
—Henri Poincaré, 1854–1912

16

An Interspecies Hope Chest

REFLECTING ON CHANGING NOTIONS OF OUR EVOLUtionary development, gorilla naturalist George Schaller quoted the *slightly* comforting thought of G. W. Corner about the human-ape: "After all, if he is an ape, he is the only ape that is debating what kind of ape he is."

That thought remained unassailable only for a time. Then, in the late 1970s, we found there was a gorilla in California who signed "Koko good . . . good gorilla . . . fine animal gorilla": debating what kind of ape *she* was.

And suddenly we were not alone. Not unless Penny Patterson or Koko the gorilla—or *somebody*—was faking or being fooled.

This debate over interspecies communication has always had high stakes, as high as those in the debate over evolution itself. It begins as a study of whether an animal can say "Gimme ice cream" but escalates to take in the nature of human destiny—and the nature of what it means to be human. It has never been a tiny subject. That's one reason emotions run so high.

In the course of researching this account, I have listened to a great deal of "He's crazy!", "They're insane!", "What a midget!", and other refreshing terms—I have no objection to people speaking their minds—in which theorists and experimenters, when off-guard, appraise colleagues with whom they have an intellectual difference.

We can say, as Anna Michel said, that they are "more competitive" than you would expect of scientists. I suspect it is different than that and deeper. Such vitriol strikes me as simply indicating that the persons involved are committed very powerfully. The battle, in its final sense, is ideological even more than it is personal.

Looking at some of the notes I had compiled, Lee said, "Scientists do slur each other, don't they?"

"Yes, they seem to."

"Are they allowed to do that?"

"Yes."

"They're going to all wind up suing each other," she predicted.

"No," I said. "They're going to all wind up writing books. They have."

"They're talking about hoaxing and fooling people and being self-deceived. It sounds like somebody could sue."

"No," I said, "I don't think so."

"What would you do if you were called to court?"

"I'd wear my strawberry and cream tie."

"Yes, but what if—"

"There's a way to settle this," I said. "But I don't think it will happen."

"What do you mean, there's a way to settle it?"

"Whether animals talk."

"How would you settle it?"

"As long as it's gone this far," I said, "I think I'd like to see a great court trial where they call the gorilla as a witness."

It was only some banter to end the day with. But later I wondered: Isn't it likely that if the interspecies communicators continue to claim as they've been claiming, something very like this will transpire? And supposing the gorilla answers; will Chomsky object?

THE "INCREDIBLE ACHIEVEMENT"

Learning language happens to each of us when we're so young we haven't been able to inform ourselves that it is almost insuperably difficult. Because we know nothing of its difficulty, somehow we go ahead and learn it. The process by which this language assimilation is accomplished defies the ability of the professional linguist to track it in its entirety.

"That a baby ever learns to talk at all is almost incredible to the person who understands the nature of the task," suggested two speech teachers, Charles T. Brown and Charles Van Riper, in a book called *Speech and Man*. "When we consider the problems encountered in being able to use meaningful language, we can hardly believe that any ordinary mortal could solve them."

Noam Chomsky seemed to deliver a body blow to theories of interspecies communication by maintaining that mortals do not solve the problem of learning language—it is solved ahead of time, before we are born, by structures placed within our brains. The researches by Chomsky implied the human child has inside a kind of Infernal (or Glorious) Machine. This special system, which linguists termed a *language acquisition device*, opens vistas for ourselves that gorillas, whales, dolphins, and chimpanzees can never aspire to. This explained how we could master the unmasterable—and the theoretic case against interspecies communication still rests on the Chomsky Objection. There is an answer to it, I think, within the examples we have looked at. No matter how great a miracle it is to learn language, apes *can* learn to talk—at least a bit—and the search for whalespeak seems well founded even if it hasn't gone very far. This is not the contradiction of Chomsky's speculation that it seems. It means merely that perhaps apes, whales, porpoises share with us an Infernal or Glorious Machine that can lead them on toward language that humans eventually will understand.

The empirical evidence for Delphinese, as Lilly called the dolphin tongue, is ambiguous, but the *deductive* evidence—the kind of judgment calls that are hard to associate with anything but advanced language use—are actually very strong. Those who imagined that Dr. Lilly's speculations were *merely* speculations, with no logical train of thought suggesting them, did him a great injustice. He found the anatomical evidence, by brain dissection, that the language ability *ought* to be there. He proceeded to record hundreds of instances in which dolphins showed an advanced reasoning power that was significantly different from that of nearly all other animals. On balance, while experimental method is still at the infant stage in conveying human language to the dolphin—and while computers like Lilly's Janus are still at the subinfant stage in making progress with the decoding of whalespeak—the deduction pattern supports Dr. Lilly. Probing experiments by Ken Norris and Earl Murchison, who are anxious not to associate themselves with "the romantic school," go much further to support the Lilly theories than they seem to guess. Louis Herman now seems to share the view that the power of cetaceans to handle abstract thought, and language along with it, is much greater than he had, at the beginning, imagined.

The principle of parsimony in science says that a "higher-level process" should not be used to account for some bewildering occurrence when a "lower-level process" would do. This principle could be construed to mean that you shouldn't assume language as an explanation for an ape's behavior when there might have been some low-level cueing to arrange the gorilla's fingers and produce the deceptive suggestion of a talking ape. Talking gorillas are certainly desirable, unless they become used-car salesmen, but anthropologist Frederick S. Hulse, tracing the origins of humankind in a book called *The Human Species*, noted, "The scientific attitude is a ruthless one. We cannot 'like to think' that something is or is not so. We can easily be, perhaps we usually are, uncertain about an answer, but we cannot accept one answer because it is more comforting or less flattering than another. . . . Spurious reasoning must be rejected even if it would lead to a conclusion we would enjoy."

It is hard to separate the wish-fulfillment element in the interspecies investigations from the actual accomplishments of the investigators. It is not as clearly understood that their critics also have exhibited a wish-fulfilling stance. Commentators like Thomas Sebeok went into each new controversy determined to show that interspecies communication *could not* be a fact. After many years of maintaining that language is strictly a human capability, they did not wish to be confronted with new evidence that it might not be so. So the idea of being ruthlessly true to whatever the facts might show could, in fact, cut both ways.

One unappreciated effect of academic scorn for the interspecies experiments has been cutting off grant money to these new-style experimenters. Dr. Lilly and Penny Patterson had to become entrepreneurs and foundation chiefs merely to keep their experiments going. The buying and training of gorillas is expensive; the decoding of whale talk is as much a financial matter as it is a matter of the intellect. The critics hold the flow of funds away from interspecies experiments ("They attack," a researcher said, "like dobermans"), and that is unfair in the light of the progress already shown.

The scorn of the academics against the new line of animal experiments has often arisen, I think, because they associate these language experiments with paranormal investigation. I have said little in this book about telepathic communication with animals as an evidence for interspecies conversation because no supportable instance can be found in the record. What we do find, in the history of telepathic interchange with either humans or animals, is duplicity and shenanigans of every sort. The history of telepathic communication is fraught with adult delinquents—and even with small children who, innocent faced and protesting their terror, have invented systems of ghost rappings that drove the adults around them crazy. Never for a moment will this kind of child, or this kind of adult, admit that what they really love is the power sense that comes from making the impossible seem real.

Arguments against telepathy can be considered on another day; but it is wrong to associate what Lilly and the ape experimenters have been up to with the paranormal. The association comes, in part, because there have been so many Sunday-supplement-type stories about telepathic communication (with no evidence to back them) and considerable confusion about the peculiar sensory equipment of the dolphin. The fact that a dolphin could be trained to find a pea under a shell seems paranormal only to those who haven't studied the acoustical techniques on which the trick is based. Radio and television—transmitting words and pictures from town to town or continent to continent—would have seemed paranormal (and impossible) until very recent times, but we are already losing the sense that there is anything magical in these phenomena. The human concept of the paranormal has been subject, throughout the last two centuries, to almost constant revision. A deeper appreciation of the special powers of whales and dolphins will probably revise it a bit more.

Should we believe Steve Sipman's wild metaphor that maybe dolphins are doing "calligraphy in the feces"? No, not likely—but porpoise formations that have the movement of the dance, whalespeak that has the movement of singing, and gorilla drumming that serves to bring out deep emotions of the inner animal should be treated as intriguing indications of animals who express themselves more intricately, and more individualistically, than usually admitted. There shouldn't be such great surprise at this. The size and complexity of the brains had indicated it. Steve's suggestion that dolphins can be seen, from the underwater Plexiglas bubble, drawing in the sand should be followed up. I keep wondering if they do and what it could mean.

A film exists, made by the Navy, which shows whales dressed in space-agey headgear nosing down into the ocean, finding a lost torpedo, clamping a lifting device onto the torpedo with their special headgear, swimming confidently away as a parachute blooms from the device they have installed. The parachute lifts the torpedo to the surface.

Earl Murchison described to me a chat he had with Dr. Lilly when the two of them watched a film like that—perhaps it was the same one I had seen—and Murchison had said, "Now, John, *that* is communication."

Dr. Lilly replied, according to Murchison, "I get the feeling the whale is way ahead of us."

In San Diego, I had spoken with Navy men familiar with the making of the film, and they had told me that not only were the whales marvelous at understanding the process the humans taught them, but the whales soon felt they understood what was wanted better than clumsy sailors did. When the sailors fouled up, the whales would get sore and fuss—in gestures the humans could understand.

It sounded to me as though Lilly had called it right. The whales *were,* at least at times, ahead of the humans who led them along.

THE LAST (COSMIC) WORD

Earlier, in speaking of the war between believers and disbelievers in the interspecies quest, I referred to "a war of the worlds." I think this is a fair description. Decriers and defenders of apetalk and whalespeak are not having a minor quibble about animal talk; they are dealing with an idea so large that, even if they do not acknowledge this, it affects our whole idea of the universe and what it is like or *should be* like.

That brings us, I think, to B. F. Skinner and Skinnerian theory in its cosmic phase.

Perhaps because he likes to have the last word in cases of animal behavior, Skinner had a short and somewhat mocking fling at the controversy between the primatologists. He came at it obliquely and seems to have been prompted by an announcement from the Rumbaughs that two chimpanzees named Sherman and Austin had "achieved the first instance of symbolic communication between nonhuman primates." Skinner and his colleagues set up a stunt for pigeons. The pigeons, whose names were Jack and Jill, supposedly mastered an exchange that was represented as comparable to what the apes were doing. *Science News* considered the Skinner effort tongue-in-cheek (they titled their report "Two Ways to Skinner Bird") and noted, "Jack was able to ask Jill about a hidden color; Jill would check the color behind a curtain and peck the answer on a coded key to Jack, who depressed a 'thank you' key that rewarded Jill with food. Jack then pressed an appropriate, color-matched key on his console and received his food." When Jack and Jill had learned to help each other in this way, Skinner announced "the first instance of such symbolic communication between nonprimates."

Skinner's pigeons were worth a smile, but the jibe was clear enough. Duane Rumbaugh noted that the pigeons needed fixed locations of keys while the keys the chimpanzees used were shifted to make sure that they understood the symbol coding itself, not just the location of some keys.

Why did Skinner bother?

As the leading behaviorist-philosopher, the one who had set the tone for several decades of mechanistic animal experiments, he may have seen trouble brewing. When chimpanzees and gorillas seemed to be exercising independent thought (free will?) to some degree, Skinner may have recognized a vague and distant threat to those theories he holds most dear.

Skinner has urged us to accept both the truth and the utility of the view that neither man nor beast has free action beyond the culture producing them. In *Beyond Freedom and Dignity,* he argues what the title implies—that all our fussing about a need for freedom can get in the way of the most desirable development of our own species. He explains that the "determination of behavior" can be shifted "from autonomous man to the environment."

"In the scientific picture," Skinner tells us, "a person is a member of a species shaped by evolutionary contingencies of survival, displaying behavioral processes which bring him under the control of the environment in which he lives, and largely under the control of a social environment which he and millions of others like him have constructed and maintained during the evolution of a culture. The direction of the controlling relation is reversed; a person does not act upon the world; the world acts upon him."

When we are wisely "beyond freedom and dignity," would we then be, in the real sense, gone? "Certainly not as a species or as an individual achiever," says Skinner. "It is the autonomous inner man who is abolished and that is a step forward."

Although Skinner is always intellectually entertaining, he expressed just what humankind tended to fear most in behaviorist philosophy: that the universe itself could be construed as a giant Skinner box in which all we do is respond to incentives. If we all come to be in fatalistic submission to a "benevolent" environment, then wouldn't we be just another animal—or just another animal act?

I don't think it is imputing too much to them to suggest that Penny Patterson's gorillas are, as nearly as anything that exists, a refutation of this ultimate Skinnerian premise. Skinner would disagree, but it appears that we have had at least a glimpse of something that might be described as "the autonomous inner gorilla": a gorilla who thinks, chooses, desires champagne, lies, schemes, surrenders, consoles, kisses, loves, makes up—and is bad again tomorrow.

It seems to me that this is what Penny has there at the Gorilla Foundation: free-will gorillas. Although their mastery of language is not great, it is growing. Because they are less devious than human beings (so far) but are equipped with our own most precious skill, language, we can see very clearly that these gorillas *are* capable of independent thought. Can the operant conditioning of which the psychologist speaks really explain a gorilla who dreams of squashing alligators?

Somehow, I doubt it.

Perhaps, if we can get all their sounds into the computer and break the code, whales will tell us more about this than gorillas can. Perhaps whales will know more about the connection between dignity and freedom than we do. Perhaps they will know more than B. F. Skinner knows.

In any case—and to be clear-cut about it—we do have apetalk. We have whalespeak only in the sense that the whales seem to *want* to speak, and their outpourings seem to have the qualities of complex conversation. We have not yet broken the code, but we seem quite close. Humpback studies imply that some types of whaletalk may prove to be more like choral singing than conversation when we finally translate it, a choral singing strongly related to mating and fertility cycles, because there are times when humpbacks are silent. But whalespeak could also be fast, shifty conversation at that, with a sexy layer to it; that's what the orca studies imply. The difference in urgency between male-male conversations and male-female conversations should be no great surprise; that is what makes the whales—well, almost anthropomorphic.

But *anthropomorphic* is a word no longer pleasing to me. In the apetalk discussion, it has been used like a blinder. Why should a humanlike chimpanzee trait be called anthropomorphic? Why don't we, instead, call the human trait "chimpomorphic"? And if gorillas are found to be like us in some ways, let us say—to be fair—that humans are "gorillaesque." If there is any trait we share with whales (and I'm not sure there is), we could say that in some ways we are "cetamorphic."

HOPE CHEST

As noted earlier, Carl Sagan turned the story of humans' seeming exclusiveness in language into a murder mystery from the primordial past. He suggested that we had killed off our near rivals in intelligence in our climb to power over the planet—a view Lilly espoused—and felt that teaching the apes to speak was a sort of constructive penance we could pay for this evolutionary crime.

That was one of the ideas, then, in the hope chest of the interspeciesist: perhaps we could use our great technological skills to repair the damage we have done to the earth and other creatures. I believe I side with that, although I don't think our chief motivation needs to be a guilt trip for the sins of pre-Stone Age man. We have crimes of the present that are quite enough to atone for. Gorillas are even now being butchered out of existence, except for those in zoos.

My own hope chest would include the proposition from James Hudnall—show to mankind the creatures talk, and then the persecutors may be willing to save them. I would like to see a talking gorilla go back to Africa, on a less melodramatic journey than the one imagined by novelist Michael Crichton, and make a headlined tour from country to country, speaking in sign language to show that it is not the devil-creature the gorilla poachers still portray it to be.

While the Lilly proposition that whales are the greatest talkers strikes me as entirely plausible, I am going to hope, while urging Lilly and the Janus computer along, that at the point where we decipher whale talk we will exchange ideas with whales only in a very occasional way. Our more important policy should be to leave them alone in their separate world lest we have a great and damaging effect upon them.

What could happen to the whales if we follow too urgently that higgledy-piggledy idea of the interspeciesist that we should sit at the whales' feet (or, rather, fins) to learn great wisdom? I think we would only prove that we couldn't manage this. Once the whale contradicted our ideas of what constitutes benevolence and joy and courage and nobility, we would likely jump up and try to teach the whale a thing or two. The interspeciesists, who often have a strong puritanical strain, would be prone to lecture the whale if it fell short of agreeing with them on what benevolence really is.

There is a ghastly story from our recent past that indicates what happens when one society invades another and absorbs a part of its language.

When the first Yankee whalers went among the Eskimos, they found great hunters who made remarkable rituals out of their forays to capture the bowhead whale. The bowhead was a luminous creature who, at night, was illuminated by thousands of glowing microscopic organisms clinging to its body. That luminescence enabled the whalers to track the bowhead through the blackest Arctic night. Harpoons enabled them to reduce the whales to a shadow of their former population and, in the process, to largely destroy the Eskimo's way of life.

The whalers imagined that they had come among a very simple people. But as the Scandinavian explorer and Arctic trader Peter Freuchen has demonstrated, the Eskimos were a people of amazing complexity. Their language reflected their intricate interdependence with each other and with the animals on whom their lives depended. In most of Western civilization, a person with a vocabulary running to thousands of different words almost qualifies as savant, but the *average* eskimo had a vocabulary of 10,000 words. There were 110 different words for snow alone to cover this phenomenon in its endless manifestations. With a sense of doom, we turn away from the *Encyclopedia Britannica*, where it is noted that wherever a village was untouched by the *tannik* (the white man), travelers to Eskimo country found "the healthiest and happiest people on earth." In his 1971 report, *The Twilight of the Primitive*, Lewis Cotlow enumerated some of the long-range effects of the *tannik* upon these "healthiest and happiest" people. He found an Alaskan native population with a suicide rate twice that of the white population of Alaska. He found Eskimo children with only half the chance of white Alaskans to survive infancy. He found that in 1966 "the average Eskimo life span was 34.5 years."

And what, meanwhile, of the glorious Eskimo language that had spurred their imaginations, promoted their great mutual cooperation, made them a people with neither prisons nor police (they had two punishments, death or disgrace, but these were not often invoked)? This language had been spoken throughout Alaska before the *tannik* came. Enormously complicated, it was learned by every child in every village. What became of it?

It was too difficult for the *tannik*, so the *tannik* made it into a language of a few hundred words. It became pidgin Eskimo, a tongue estimated by the *Britannica* to commonly run to 500 words. Language was smashed, culture was smashed, spirit was smashed. Cotlow quoted an Alaskan native who said, "Eskimos feel inferior. They feel very inferior and sorry for themselves."

The greatest hope, in my own interspecies hope chest, is that when we *do* decipher whale and dolphin, we have the strength only to eavesdrop—to leave them alone, not fix them up.

The final irony, the final demolishment, of the interspecies dream would be this:

We teach the dolphin pidgin Dolphin.

We teach the whale pidgin Whale.

We give them a pidgin, mutilated, miniaturized, human-centered, humanesque existence. We tear their culture apart. We wreck them. We follow the path we have followed with the primitive peoples.

Our adventure into interspecies communication can turn out well only if we let it fan the great movement for a conservation of all that lives, walks, crawls, and swims, and if we overcome our persistent human tendency to destroy all kinds of living except our own and to call our smashing of other worlds a "civilizing influence."

POSTSCRIPT

The Infection of Language

WHILE DEALING WITH PROFESSIONALS IN SYNTAX and linguistics, so much rumpus is raised over the most obvious and primitive aspects of language that somehow the talk never gets around to what is most mysterious, and probably most important, in the words we speak: their power to infiltrate where least expected.

The process by which new words and new figures of speech are instilled in our personality is bewildering. How do we choose what we choose? And why?

Something about language, especially in its seemingly careless forms, is "catching," deeper than all grammar and syntax of whatever kind. Distracted but wanting to say something when my wife appears on the staircase and my hands are still on the typewriter, I mutter absently, "How yuh come on theah?" Only when she complains do I hear what I have said and realize that this greeting has somehow stuck in my mind from years and years ago when Mr. Uzzle, the wiry little city manager of a tiny southern town, would invariably greet his callers with the cheery cry, "How yuh come on theah?"—almost a foreign tongue, as Mr. Uzzle intoned it, but a sound so friendly and charming that it follows me still.

My wife came to me complete with an argot—wry, sophisticated, unexpected—picked up from Greenwich Village to Hollywood, and she could round off a formidable discussion of Van

Gogh by crying "Damn tootin'!" to emphasize a point. I know that I married her partly because of the way she said "Damn tootin'!" although why these words should have this mesmerizing charm I have no idea.

Sorting back through the language that has stayed with me, I can sometimes identify its donors. Eloquence has little to do with it—I have known eloquent people by the score who left no mark on me or on anyone else. But all who ever met him were affected by the blurted, testy speaking style of Mr. Larson, a farmer who moved into town at Kalispell, Montana, and set up a combination grocery store and soda fountain on the far corner of Main Street. His store was little more than a rundown shack but it had a fatal attraction for the high school crowd, and so did Mr. Larson.

Nobody spoke more commonly, was less affected by the jargon the high school kids strewed around town, or made less effort to doctor his speech into some pleasing blend of comedy and grace. How, then, could Mr. Larson affect us when our linguistic woods were full of every kind of jazz and bebop?

Now I really don't know. I suppose it had to do with a certain air Mr. Larson had of being on to great things we smartcracking high school kids didn't know and could probably never learn.

"Say, young man!" he would challenge. "You think you are so smart! Tell me, young man, who was Ole Bull? Terrible, terrible, Crail! You didn't know who Ole Bull was? Young man! Where's your education?" Turning derisively to the bunch at the punchboard, Mr. Larson would make a great din about such ignorance. "Say! The great violinist Ole Bull, who could make bullfrogs jump through a window—Crail never heard of him!" Shaking his head, exposing his bad teeth, he would clump by me, muttering, "Young man, I'm surprised at you!"

Often the knowledge that came to Mr. Larson happened to be great national secrets—inside stuff from the topmost channels. Perhaps he could divulge it, perhaps he couldn't. The repositor of such secrets must guard himself carefully. Oh, it was very hush-hush, the sort of things Mr. Larson knew. He could be depended on, though, to give us a hint—and to laugh, in short barks, at our astounding lack of knowledge.

"Say! Young man! How do you like your President Roosevelt now that it's come out?"

"FDR, Larson. We who knew him best called him FDR."

"The biggest Roosevelt scandal ever, Crail didn't even hear about it! Still in the dark! Young man, I'm ashamed of you!"

We could tease him by pretending to ignore his hints but Mr. Larson was an insistent devil. His crowing notes of triumph could drive you mad.

"Why didn't they open Roosevelt's coffin—tell me that, young man! What were they *hiding*? Oh, you didn't know about it, did you? The sealed coffin! Crail didn't even hear about it! Nobody can see the body! You think that's the way we treat our presidents? You think they didn't put Lenin in a glass coffin?"

"All right, Mr. Larson. *Why wouldn't they open Roosevelt's coffin?*" (The idea was to ignore him, but this was hard, and if you got right down and begged, he'd 'fess up.)

"Roosevelt was shot by his brother-in-law as a sex pervert! You think they're going to show the bullet wounds for the whole world to know? Young man, I'm surprised at you! No education!"

There were many Larsonisms, a dictionary of them. Even now, when I run into friends from those days, sometimes we begin, "Say, young man! You think you are so smart . . .!"

Anyone with a strong idiosyncratic way of speaking—something so personal they could put a patent on it—can forever affect the way we speak. I surely didn't expect or plan to learn anything about talking from gorillas, but I am easily influenced, as perhaps you are, and it turns out their style is infectious.

As I lived with gorilla language for a while and, without intending it, I developed a new way of waking up in the morning.

"Ted, are you awake?"

"Think squash, Lee."

"Yes, I know, but you better get up. Didn't you have an appointment? I'll cook you some sausage."

"That good, Lee."

Twenty minutes later:

"Here's the sausage. I'm worried about the painting I'm working on."

"Alligator."

"You think you're a gorilla now? When are you going to finish the book?"

"Know, Lee."

"Look, I could damn tootin' get tired of this in a hurry. There are some things I have to talk to you about before I leave here."

"Think squash."

"I'm going to take your car keys, okay?"

"Strangle."

"Look, you're not funny. I want the keys. I don't want to go back upstairs."

"Think go, Lee."

"Can I have the car keys or not?"

"Strangle."

"Okay, you're a gorilla monkey and I want the car keys. . . . Thanks."

"Teddy elephant devil."

"Teddy not elephant devil. Teddy dirty bad toilet."

Like all forms of language that seem out of key with what you're used to, gorilla talk sets up echoes. I can almost guarantee that, if you'll put some study into the muddled but memorable gorilla dialogues that appear in this book, you'll develop an urge to try, sometimes, to see things as a gorilla sees them. I'm not just trying to indicate that things are cute at Teddy's house. A few seconds of gorilla talk wears thin, then thinner. But I pursue it because, having gotten close to a language that is not, like Russian, a pain to be conversant in, I find the shorthand of it—the easy way that gorillas find to say something—has an appeal on a sleepy morning.

Mr. Larson's rattled reasoning produced one of the lowest forms of talk in Kalispell, Montana. Low but compelling. The town's cockiest quarterback would come in, fresh from his touchdown run, and Mr. Larson would begin, "Say, young man!" and have him in shreds in a moment with recollections of the *real* football players like the Bongo Whistler. "You never heard of the Bongo Whistler? I'm ashamed of you!"

Larson is dead. His language lives after him.

I don't know how soon gorilla language will leave me or if I will go on to make too much of it. Gorilla jabberings have the advantage of being almost clear and still mysterious—each word lurks somewhere with a suggestion of gorilla secrets and gorilla knowledge that you and I can at best barely glimpse. The gorilla responds, but he is in a different place than we are. "Alligator strangle" means so little that it seems to mean a lot. Or—let's not underestimate a gorilla—how little does it mean?

Dead for years, Mr. Larson still jeers at me at times as I lie in bed at night.

"Say, young man! You think you went off to the city and got smart! Is that it?"

"Damn tootin', Mr. Larson."

"Young man, tell me! Who was the Hollering Canute? Ignorant, young man! Ignorant!"

"Strangle, Mr. Larson."

"Young man—!"

"Strangle! Squash!"

And at long last, I believe I have the best of Mr. Larson. The gorillas have come to my rescue.

Index